Radar Workbook

— Problems and Answers in Marine Radar Operations

by
David Burch
and Larry Brandt

STARPATH

Illustrations and design by Tobias Burch and Larry Brandt.

Cover includes a Nobeltec radar image overlaid on a chart in Visual Navigation Suite ECS, with a screen shot from a Furuno radar.

Manufactured in the United States of America

Published by Starpath Publications, 3050 NW 63rd St., Seattle, WA 98107
www.starpathpublications.com
01.06.11

ISBN 978-0-914025-15-3

TABLE OF CONTENTS

Foreword

This Workbook is designed to be used in classroom or online courses in radar, or for individual study outside of the classroom. The Lesson structure follows that used by several schools in the US, based on the background reader *Radar for Mariners* (McGraw Hill, 2005) by David Burch.

An annotated slide show presentation and *Instructor's Guide* are also available from Starpath Publications for schools using the book.

The Starpath Radar Trainer PC radar simulator is recommended for classroom demonstrations or personal study. It is available at navigation outlets worldwide or online at www.starpath.com.

The book is intended for all radar users, recreational and professional. It covers the use of radar for chart navigation, blind pilotage, and for collision avoidance.

The Appendix on advanced radar plotting has been added for professional mariners who seek more practice on interpreting ARPA output by working out the vector solutions themselves. The ability to manually interpret the radar interactions seen on the screen independent of electronic solutions is in keeping with the fundamental tenet of good navigation and seamanship that we should not rely on any one aid alone.

— Lesson 1 —
How Radar Works

Terms and Concepts

- Radar target
- Open array and radome
- Horizontal beam width
- Range and range rings
- Heading line
- Cursor
- EBL and VRM
- Relative bearing
- Head-up, North-up, and Course-up
- Brilliance and Gain
- Portable range scale

Class Notes:

Lesson 1 Questions

1-1. When you first turn on a radar unit, what is the typical warm-up time before you can transmit?

❑ (A) None required

❑ (B) 30 seconds

❑ (C) 1 to 2 minutes

❑ (D) 10 to 12 minutes

1-2. Your heading is 050 M. The EBL is set on a target bearing 325 R. The magnetic bearing to the target is?

❑ (A) 325 M

❑ (B) 050 M

❑ (C) 275 M

❑ (D) 015 M

1-3. The typical horizontal beam width of an 18-inch wide antenna is about?

❑ (A) 2°

❑ (B) 6°

❑ (C) 10°

❑ (D) 18°

1-4. Your scanner is 16 ft high, a large target is 36 feet high, how far off might you first see this target on radar?

❑ (A) 6 nmi

❑ (B) 10 nmi

❑ (C) 24 nmi

❑ (D) 36 nmi

1-5. When you turn your vessel, your heading line rotates on the radar screen. What display mode are you in?

❑ (A) Head-up only

❑ (B) Course-up or Head-up

❑ (C) Course-up or North-up

❑ (D) North-up only

1-6. At night, for the best picture you usually have to....

❑ (A) Turn the Brilliance up

❑ (B) Turn the Brilliance down

❑ (C) Turn the Gain up

❑ (D) Turn the Gain down

Lesson 1. How Radar Works — Points to Ponder

Background reading is RFM chapters 1, 2, and 3, with parts of 7 and 8.

Review the background reading and class notes to formulate your understanding of these points, as if you were preparing to explain the concepts to a friend. Strive to distill your thoughts to two or three key sentences.

1. What are the characteristics of a good radar target?

_____

_____

_____

2. What is the relationship between antenna width and radar beam width?

_____

_____

_____

3. Explain the difference between Brilliance and Gain controls.

_____

_____

_____

4. List three ways to measure the range to a radar target with their merits.

_____

_____

_____

5. List three uses of the cursor on a radar screen.

_____

_____

_____

6. Explain the differences between Head-up and North-up display modes.

_____

_____

_____

7. Explain the difference between North-up and Course up display modes.

_____

_____

_____

8. List three major specifications of a radar system.

_____

_____

_____

— Lesson 2 —
Radar Operation

Terms and Concepts

- Anti clutter sea and STC
- Anti clutter rain and FTC
- Tuning control
- Plots and wakes
- Electronic Range and Bearing Line (ERBL)
- Pulse length
- Pulse repetition Interval
- 6-minute rule

Class Notes:

_____
_____
_____
_____
_____
_____
_____
_____
_____
_____
_____
_____
_____
_____
_____
_____
_____
_____
_____
_____
_____
_____

Lesson 2 Questions

2-1. The most appropriate initial radar-picture adjustment sequence is usually...

❑ (A) Brilliance, Gain, Range

❑ (B) Gain, Range, Brilliance

❑ (C) Brilliance, Range, Gain

❑ (D) Range, Gain, Brilliance

2-2. The normal setting of the Anti-clutter Sea control is

❑ (A) Full on till sea clutter reduction is called for

❑ (B) Full off till needed

❑ (C) Half scale till adjustment is called for

❑ (D) Set to match the Anti-clutter Rain control

2-3. At 1230 a target has range 8.5 mi, bearing 050 R and at 1236 the range is 7.5 mi at bearing 050 R.

❑ (A) Its speed of relative motion is 10 kts

❑ (B) Its speed of relative motion is greater than 10 kts, depending on our speed

❑ (C) Its speed of relative motion is less than 10 kts, depending on our speed

❑ (D) Its speed of relative motion cannot be known without knowing our speed

2-4. A 3-minute target wake is 1.2 miles long. What is its speed of relative motion?

❑ (A) 4.0 kts

❑ (B) 6.0 kts

❑ (C) 12.0 kts

❑ (D) 24.0 kts

2-5. A short pulse length would typically be about how long...

❑ (A) 3 yards

❑ (B) 30 yards

❑ (C) 100 yards

❑ (D) 300 yards

2-6. To recognize a tug and tow as two targets rather than one, you would be best to...

❑ (A) Turn up the Gain

❑ (B) Turn down the Gain

❑ (C) Switch to the lowest range that still shows the target

❑ (D) Use the Offset to move the target to the outer edge of the screen

Lesson 2. Radar Operation — Points to Ponder

Background reading is RFM chapters 1, 2, and 3, with parts of 7 and 8.

Review the background reading and class notes to formulate your understanding of these points, as if you were preparing to explain the concepts to a friend. Strive to distill your thoughts to two or three key sentences.

1. Describe the structure of a radar beam—its physical dimensions and electronic frequencies.

_____

_____

_____

2. Rain clutter versus sea clutter... what are these?

_____

_____

_____

3. Explain the role of the Anti-clutter Rain control and how it can be used besides rain clutter control.

_____

_____

_____

4. Explain the role of the Anti-clutter Sea control and how it can be used beyond clutter control.

_____

_____

_____

5. List reasons for changing pulse length (when you have the option).

_____

_____

_____

6. Explain the difference between an EBL and an ERBL.

_____

_____

_____

7. Explain the role of target trails and wakes.

_____

_____

_____

8. Explain how you would measure how fast a target is moving across the radar screen.

_____

_____

_____

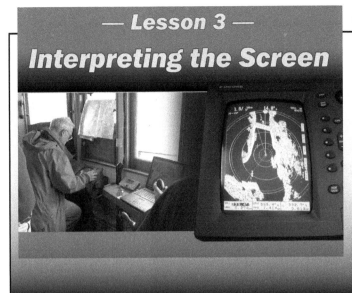

— Lesson 3 —
Interpreting the Screen

Terms and Concepts

- Horizontal shadows
- Vertical shadows
- Effect of beam width on display
- Use of ECS for radar orientation
- Side lobe interference
- Radar to radar interference

Class Notes:

Lesson 3 Questions

3-1. You can see a dinghy just 200 yards away, but it does not show up on the radar. What might cause this?

- ❑ (A) Interference Rejection has been left on
- ❑ (B) Anti-clutter Sea set too high
- ❑ (C) Anti-clutter Rain set too high
- ❑ (D) Your antenna is too high

3-2. You are travelling along a shoreline and see a large target on the radar just off the shoreline, but none is there. What are you likely seeing?

- ❑ (A) The effect of radar shadows on a shoreline feature
- ❑ (B) Side lobe interference
- ❑ (C) An effect of your beam width
- ❑ (D) An effect of your pulse length

3-3. As you pass an anchored ship with its radar still running your radar image of the ship smears out around the screen along the range to the ship. What is most likely taking place?

- ❑ (A) Your radar signal is bouncing back and forth between you and the ship causing multiple targets
- ❑ (B) The ship's radar beam is saturating your antenna and distorting its image
- ❑ (C) You are experiencing side lobe interference and need to turn down the Gain
- ❑ (D) You have reached the minimum range of your radar unit and this cannot be avoided

3-4. When you use the EBL to measure the bearing to the left-hand side of a small island, you can expect that bearing measurement to be...

- ❑ (A) About right if your compass is adjusted
- ❑ (B) Slightly too high
- ❑ (C) Slightly too low
- ❑ (D) This depends on whether we are in Head-up or North-up display mode

3-5. Transient spiraling streaks or fields of dashes on your radar screen are caused by

- ❑ (A) Poor tuning
- ❑ (B) Side lobe interference
- ❑ (C) Power line interference
- ❑ (D) Radar to radar interference

Lesson 3. Interpreting the Radar Screen — Points to Ponder

Background reading is RFM chapters 1, 2, and 3, with parts of 7 and 8.

Review the background reading and class notes to formulate your understanding of these points, as if you were preparing to explain the concepts to a friend. Strive to distill your thoughts to two or three key sentences.

1. Describe the steps you would go through when you first look at the radar to get oriented with the picture it shows.

_____

_____

_____

2. Give a few examples of how horizontal beam width distorts the radar picture.

_____

_____

_____

3. Describe what determines the minimum echo (pip) size we can see on the radar screen. (Think of radial vs. azimuthal dimensions of the target image on the radar screen .)

_____

_____

_____

4. Explain the distinction between vertical shadows and horizontal shadows as seen on the radar.

_____

_____

_____

5. We see on the radar what looks like a ship anchored along a steep shore a couple hundred yards on our right but we can clearly see that no ship it there. What are we seeing?

_____

_____

_____

6. Summarize the role of electronic charting options when it comes to understanding the radar picture.

_____

_____

_____

7. Describe two types of common radar interference and how to remove them.

_____

_____

_____

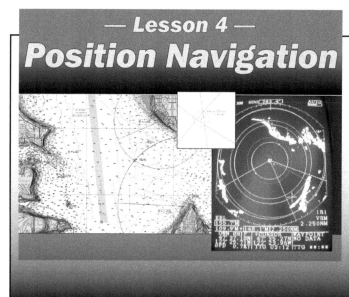

— Lesson 4 —
Position Navigation

Terms and Concepts

- Confirm GPS with radar
- Range and bearing fix
- Range and tangents Fix
- Multiple range fix
- Pros and cons of fix methods

Class Notes:

Lesson 4 Questions

4-1. One of the best possible radar targets for navigation would be

❑ (A) A SART signal

❑ (B) A large light house on the shoreline

❑ (C) A RACON signal

❑ (D) A buoy with a radar reflector on it

4-2. A range and bearing fix to a single target could be described as

❑ (A) Fast and precise

❑ (B) Fast but not as accurate as other methods

❑ (C) The preferred type of fix because only one target is required

❑ (D) The best type of fix if you have a heading sensor

4-3. The most accurate radar fix is made from.

❑ (A) Range and bearing to a single target

❑ (B) Crossed bearings from two or more targets

❑ (C) Intersecting ranges from 2 or more targets

❑ (D) Bearings to both sides and a range to the center of an isolated target

4-4. The quickest way to check your GPS position with the radar is

❑ (A) Range and bearing to a single target

❑ (B) Two crossed bearings

❑ (C) Two crossed ranges

❑ (D) A range to one target and a bearing to another target

4-5. A standard tool on electronic chart displays that helps the most with understanding the radar image is

❑ (A) The range and bearing tool in the chart program

❑ (B) A split chart screen display showing large and small scales side by side

❑ (C) The position projection vector on the vessel icon

❑ (D) Setting range rings on the vessel icon to match those on the radar

4-6. The vertical divergence of a radar beam well away from the vessel is about

❑ (A) 25 to 30 degrees, more or less independent of antenna size

❑ (B) 2 to 8 degrees depending on antenna width

❑ (C) Varies from 5 to 30 degrees depending on radar model

❑ (D) Stays at a constant profile without diverging at all in the vertical plane

Lesson 4. Position Navigation — Points to Ponder

Background reading is RFM chapters 4 and 10.

Review the background reading and class notes to formulate your understanding of these points, as if you were preparing to explain the concepts to a friend. Strive to distill your thoughts to two or three key sentences.

1. Explain how radar can be used for a quick check of the GPS position?

_____

_____

_____

2. Describe several ways to get an actual chart position from radar observations?

_____

_____

_____

3. Why are radar ranges more accurate than radar bearings?

_____

_____

_____

4. What is the most accurate type of radar fix and how do you carry it out?

_____

_____

_____

5. What are a few circumstances where a radar fix might be better than a GPS fix?

_____

_____

_____

6. How do you answer the question "Is it safe to be around the radar scanner?"

_____

_____

_____

— Lesson 5 —
Piloting

Terms and Concepts

- Rounding a corner
- Finding an entrance
- Following a shoreline
- Maintaining channel position
- Heading to the right channel
- Anchoring
- Effect of currents on radar observations

Class Notes:

Lesson 5 Questions

5-1. Which tool would help the most in rounding an island at a fixed distance off the shoreline?

- ❑ (A) Heading line
- ❑ (B) EBL
- ❑ (C) VRM
- ❑ (D) ERBL

5-2. Which tool would help choosing the course to an island on the starboard bow before we turn?

- ❑ (A) Heading line
- ❑ (B) EBL
- ❑ (C) VRM
- ❑ (D) ERBL

5-3. Which tool would be the most valuable for checking that our course is parallel to a shoreline?

- ❑ (A) Heading line
- ❑ (B) EBL
- ❑ (C) VRM
- ❑ (D) ERBL

5-4. Which tool would help the most to measure the width of a channel several miles in front of us?

- ❑ (A) Heading line
- ❑ (B) EBL
- ❑ (C) VRM
- ❑ (D) ERBL

5-5. Which tool might help the most to position our boat in the center of a small cove?

- ❑ (A) Heading line
- ❑ (B) EBL
- ❑ (C) VRM
- ❑ (D) ERBL

5-6. Which tool would we use to measure the bearing between two buoys, both about a mile from us?

- ❑ (A) Heading line
- ❑ (B) EBL
- ❑ (C) VRM
- ❑ (D) ERBL

Lesson 5. Radar Piloting — Points to Ponder

Background reading is RFM chapter 5.

Review the background reading and class notes to formulate your understanding of these points, as if you were preparing to explain the concepts to a friend. Strive to distill your thoughts to two or three key sentences.

1. How far does a radar see—sometimes referred to as the geographic range of the radar—and what does it depend upon?

_____

_____

_____

2. How does the geographic range of a radar installation relate to the maximum range setting on the instrument?

_____

_____

_____

3. Describe how you might use radar to round a corner maintaining a safe distance off the shoreline.

_____

_____

_____

4. Describe two uses of the heading line for radar piloting.

_____

_____

_____

5. Describe at least one other standard visual piloting technique (done without a chart) that can be done well with radar.

_____

_____

_____

6. Outline some of the ways radar can help with anchoring.

_____

_____

_____

7. You see a buoy moving away from the heading line, but you are holding a steady heading. What is going on?

_____

_____

_____

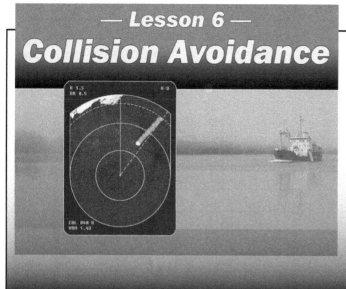

Terms and Concepts

- Closest Point of Approach (CPA)
- Parallel Targets
- Up-screen and down-screen targets
- Speed of relative motion (SRM)
- Direction of relative motion (DRM)
- Buoy trails
- Relative motion diagram

Class Notes:

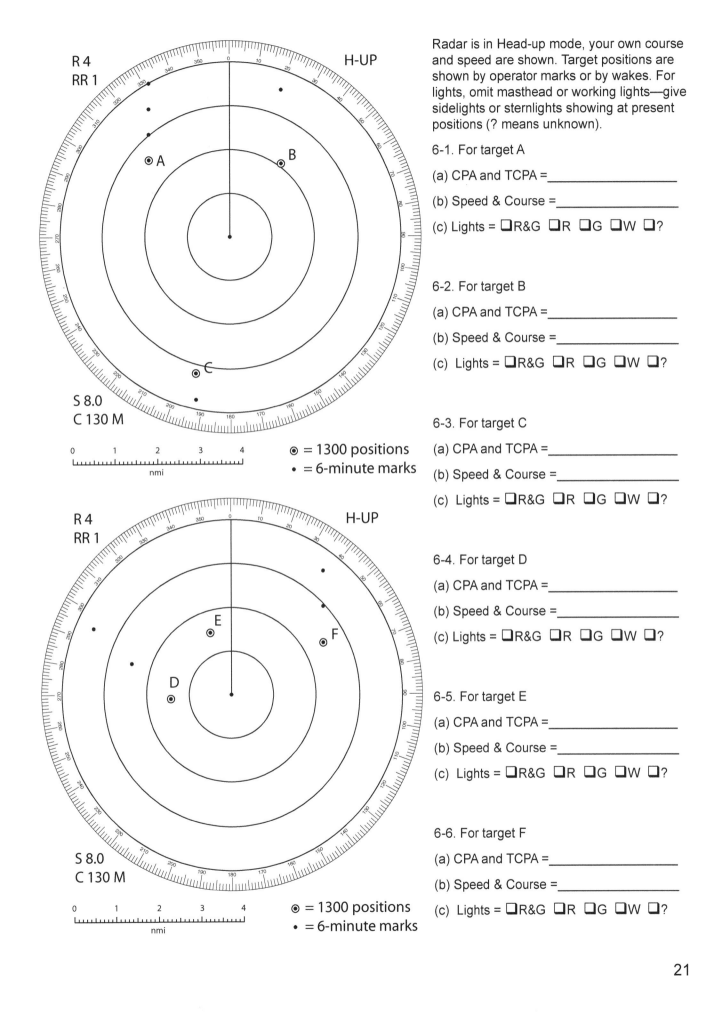

Radar is in Head-up mode, your own course and speed are shown. Target positions are shown by operator marks or by wakes. For lights, omit masthead or working lights—give sidelights or sternlights showing at present positions (? means unknown).

6-1. For target A

(a) CPA and TCPA =_____

(b) Speed & Course =_____

(c) Lights = ☐R&G ☐R ☐G ☐W ☐?

6-2. For target B

(a) CPA and TCPA =_____

(b) Speed & Course =_____

(c) Lights = ☐R&G ☐R ☐G ☐W ☐?

6-3. For target C

(a) CPA and TCPA =_____

(b) Speed & Course =_____

(c) Lights = ☐R&G ☐R ☐G ☐W ☐?

6-4. For target D

(a) CPA and TCPA =_____

(b) Speed & Course =_____

(c) Lights = ☐R&G ☐R ☐G ☐W ☐?

6-5. For target E

(a) CPA and TCPA =_____

(b) Speed & Course =_____

(c) Lights = ☐R&G ☐R ☐G ☐W ☐?

6-6. For target F

(a) CPA and TCPA =_____

(b) Speed & Course =_____

(c) Lights = ☐R&G ☐R ☐G ☐W ☐?

⊙ = 1300 positions
• = 6-minute marks

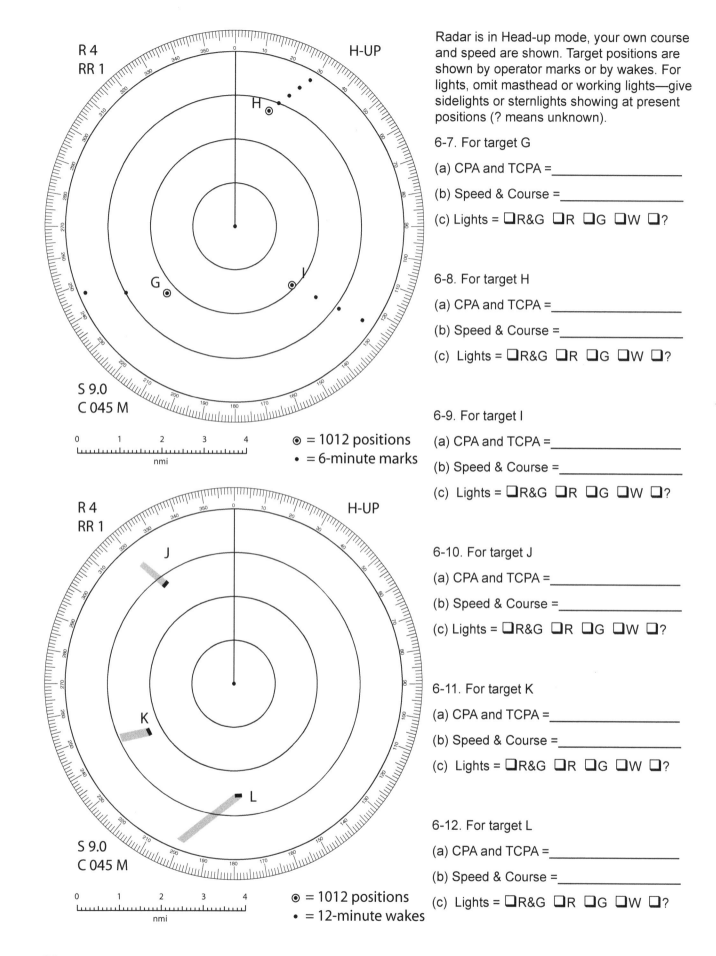

Radar is in Head-up mode, your own course and speed are shown. Target positions are shown by operator marks or by wakes. For lights, omit masthead or working lights—give sidelights or sternlights showing at present positions (? means unknown).

6-7. For target G

(a) CPA and TCPA =_____

(b) Speed & Course =_____

(c) Lights = ☐R&G ☐R ☐G ☐W ☐?

6-8. For target H

(a) CPA and TCPA =_____

(b) Speed & Course =_____

(c) Lights = ☐R&G ☐R ☐G ☐W ☐?

6-9. For target I

(a) CPA and TCPA =_____

(b) Speed & Course =_____

(c) Lights = ☐R&G ☐R ☐G ☐W ☐?

6-10. For target J

(a) CPA and TCPA =_____

(b) Speed & Course =_____

(c) Lights = ☐R&G ☐R ☐G ☐W ☐?

6-11. For target K

(a) CPA and TCPA =_____

(b) Speed & Course =_____

(c) Lights = ☐R&G ☐R ☐G ☐W ☐?

6-12. For target L

(a) CPA and TCPA =_____

(b) Speed & Course =_____

(c) Lights = ☐R&G ☐R ☐G ☐W ☐?

Upper radar display:
R 4
RR 1
H-UP
S 9.0
C 045 M

⊙ = 1012 positions
• = 6-minute marks

0 1 2 3 4 nmi

Lower radar display:
R 4
RR 1
H-UP
S 9.0
C 045 M

⊙ = 1012 positions
• = 12-minute wakes

0 1 2 3 4 nmi

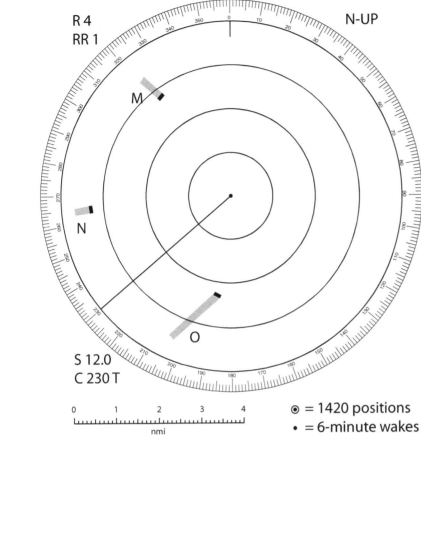

R 4
RR 1

N-UP

M

N

S 12.0
C 230 T

0   1   2   3   4
nmi

⊙ = 1420 positions
• = 6-minute wakes

Radar is in North-up mode, your own course and speed are shown. Target positions are shown by operator marks or by wakes. For lights, omit masthead or working lights—give sidelights or sternlights showing at present positions (? means unknown).

6-13. For target M

(a) CPA and TCPA =_____

(b) Speed & Course =_____

(c) Lights = ☐R&G ☐R ☐G ☐W ☐?

6-14. For target N

(a) CPA and TCPA =_____

(b) Speed & Course =_____

(c) Lights = ☐R&G ☐R ☐G ☐W ☐?

6-15. For target O

(a) CPA and TCPA =_____

(b) Speed & Course =_____

(c) Lights = ☐R&G ☐R ☐G ☐W ☐?

Lesson 6. Collision Avoidance, evaluating the risk of collision — Points to Ponder

Background reading is RFM chapters 6 and 11.

Review the background reading and class notes to formulate your understanding of these points, as if you were preparing to explain the concepts to a friend. Strive to distill your thoughts to two or three key sentences.

1. Describe the "call to action" — first things you do when first detecting a new radar target on the screen.

_____

_____

_____

2. Why can it be difficult to predict a reliable CPA when the target is first seen on the radar screen?

_____

_____

_____

3. When we measure and predict a CPA, what are the basic assumptions we are making?

_____

_____

_____

4. If a target vessel slows down, how would that appear to us on the radar screen?

_____

_____

_____

5. There is a target headed straight toward us on the screen from a bearing of 045 R (DRM=225 R), leaving a nice straight wake behind it, and then we slow down significantly. Describe what happens to that target's subsequent DRM and associated wake.

_____

_____

_____

6. We are passing a buoy that shows a well defined 3-minute wake on our radar, and we also see on the screen two targets with wakes twice as long as the buoy's trail, one is moving straight up-screen, the other straight down-screen. Describe the course and speed of these two targets relative to our own.

_____

_____

_____

7. In your own words, describe the process of determining a target vessel's true course and speed based on its 6-minute wake on the radar screen.

_____

_____

_____

8. List the cautions we must mind to when taking advantage of a "mini" ARPA option.

_____

_____

_____

— Lesson 7 —
Radar and The Navigation Rules

Terms and Concepts

- Ordinary practice of seamen
- Special circumstances
- Restricted visibility
- Close quarters
- Proper look-out
- Safe speed
- Systematic observations
- Turns large enough to be seen on radar
- Maneuvering diagram

Class Notes:

Lesson 7 Questions

7-1. What might be a guideline for a minimum turn that would be detected readily on another vessel's radar?

- [ ] (A) 15°
- [ ] (B) 30°
- [ ] (C) 60°
- [ ] (D) 90°

7-2. What is one difference between overtaking in the fog versus in clear weather?

- [ ] (A) The preferred side to pass on is optional in clear weather but not in the fog
- [ ] (B) In the fog we do not always hold course and speed when being overtaken
- [ ] (C) A sailing vessel has right of way when overtaking in the fog, but not in clear weather
- [ ] (D) Overtaking is one situation that is identical in fog and clear weather

7-3. What is the best way to summarize a sailing vessel's right of way in the fog

- [ ] (A) Sail has more rights in the fog than in clear weather
- [ ] (B) Sail has less rights in the fog than in clear weather
- [ ] (C) Sail has the same level of right of way in the fog as in clear weather
- [ ] (D) There is no right of way in the fog, not for power nor for sail

7-4. A moving radar target (not in sight) is converging in the fog on your port bow, what should you do?

- [ ] (A) Turn right
- [ ] (B) Turn left
- [ ] (C) Hold course and speed
- [ ] (D) Slow to bare steerage

7-5. A moving radar target (not in sight) is converging in the fog on your starboard bow, what should you do?

- [ ] (A) Turn right
- [ ] (B) Turn left
- [ ] (C) Hold course and speed
- [ ] (D) Slow to bare steerage

7-6. A moving radar target (not in sight) is converging in the fog on your port quarter, what should you do?

- [ ] (A) Turn right
- [ ] (B) Turn left
- [ ] (C) Hold course and speed
- [ ] (D) Slow to bare steerage

Lesson 7. Radar and the Navigation Rules — Points to Ponder

Background reading is RFM chapter 12.

Review the background reading and class notes to formulate your understanding of these points, as if you were preparing to explain the concepts to a friend. Strive to distill your thoughts to two or three key sentences.

1. Summarize the important distinction between requiring the use of radar as part of a proper look-out (Rule 5) and the requirement to use radar to evaluate risk of collision (Rule 7).

_____

_____

_____

2. Summarize the radar usage requirements with regard to choosing a safe speed (Rule 6b).

_____

_____

_____

3. What do sailors have to learn from the wording of these safe speed rules on radar, even though the rules were obviously directed more toward larger power-driven vessels?

_____

_____

_____

4. What is meant by "systematic (radar) observations" required in Rule 7 on Risk of Collision?

_____

_____

_____

5. What do the Rules mean when they tell us not to rely on "scanty radar information" in Rule 7 when evaluating risk of collision?

_____

_____

_____

6. Give a one short sentence statement of what you do when you see a converging target on the radar, but cannot see it visually?

_____

_____

_____

7. In your own words, explain the concept of "close quarters."

_____

_____

_____

8. What is an almost guaranteed way to not be involved in a collision at sea—staying at home is not the answer!

_____

_____

_____

ANSWERS

Lesson 1. How Radar Works

1-1. (C) 1 to 2 minutes

1-2. (D) 015 M

1-3. (B) 6°

1-4. (B) 12 nmi

1-5. (C) Course-up or North-up

1-6. (B) Turn the Brilliance down

Lesson 1. Points to Ponder

1. Big, close, tall, and with structure large compared to 3 cm. Note that metal vs. wood or fiberglass etc is not as big a factor as often thought.

2. The wider the antenna, the narrower the beam. The narrower the beam (small HBW) the better the images. Sample is 18 inch radar has HBW of about 6°.

3. Brilliance controls brightness of the images on the video display. Gain controls the sensitivity of the radar receiver.

4. Estimate from RR spacing: fast, no hands, but least accurate. Cursor on target: fast and accurate, but no permanent mark. VRM: accurate, but takes an adjustment, but can be left on the screen for reference.

5. Measure range and bearing. Set EBL and VRM (in some models). Select center of ERBL. Select center of Offset. Set alarm zones in some models.

6. North-up stabilized and heading line rotates to match your turn, targets remain fixed; top of the screen is always 000 T; top of the heading line is your true heading. Head-up unstabilized, heading line always straight up, top of screen reads 000 R which always represents your present heading.

7. In North-up the top of the screen is always 000 T, whereas in Course up the top of the screen is set on demand by the user to present heading. Else they behave the same. Heading line rotates to match your turns in both modes. North-up allows for direct comparison of chart and radar screen orientations.

8. Maximum range, power output, antenna width. These are usually related in that lower-end units have all three lower values than those of a higher-end unit.

Lesson 2. Radar Operation

2-1. (C) Brilliance, Range, Gain

2-2. (B) Full off till needed

2-3. (A) Its speed of relative motion is 10 kts

2-4. (D) 24.0 kts

2-5. (B) 30 yards

2-6. (C) Switch to the lowest range that still shows the target

Lesson 2. Points to Ponder

1. The "beam" some distance from the boat is a sequence of isolated pulses of microwave energy that diverge a few degrees horizontally (HBW) and about 30° vertically. The length (leading edge to trailing edge) of the pulse varies from about 30 to 100 yards depending on Range selected. The microwave energy within the pulse has a wavelength of 3 cm. The pulses are well separated in space and time, one going out and reflected back long before the next one is sent out. At some miles from the boat, the pulse is essentially a vertical wall of energy with a thickness of the pulse length. A distant target is hit by about 50 pulses each time the beam sweeps by it.

2. We distinguish the word "clutter" from the Anti-clutter control that removes it. Sea clutter is radar beam reflections from nearby waves. Rain clutter is radar beam reflections from precipitation (rain, snow, hail). Big seas, big sea clutter. Heavy rain, big rain clutter. There are separate controls on the radar to diminish each of these.

3. AC Rain is used to reduce rain clutter or to "clip" all echoes for a sharper image in some circumstances.

4. AC Sea is used to reduce sea clutter. It works by adding extra Gain reduction at lower ranges. Its effect increases in strength and range as you increase the control, but has no effect at medium or longer ranges. It can also be used for extra Gain reduction at close ranges as in a marina or when anchoring in close quarters.

5. To look for weak targets you want the longest pulse length. To get the highest resolution to separate targets at the same bearing you want the shortest pulse length. Usually these are changed automatically when you change the Range. Short pulses on low ranges, long pulses on higher ranges, to get more power sent out the farther the beam has to travel.

ANSWERS

6. EBL always measures bearings from your location at the base of the heading line, i.e. boat to target. ERBL measures bearings from an offset location on the screen that you select, independent of the heading line location, i.e. target to target.

7. Target wakes show the past locations of the targets. Without otherwise marking the screen by hand, this is how you tell which way a target is moving and how fast it is moving. Such wakes or similar trails or hand marks are crucial to evaluating the risk of collision.

8. Measure the length of the wake relative to the range ring spacing and divide that by the time the wake has accumulated. Or mark the target at two different times and do the same (speed = distance between marks divided by time between marks.) You are measuring the speed of relative motion this way, not the target's true speed.

Lesson 3. Interpreting the Radar Screen

3-1. (B) Anti-clutter Sea set too high

3-2. (A) The effect of radar shadows on a shoreline feature

3-3. (C) You are experiencing side lobe interference and need to turn down the Gain

3-4. (C) Slightly too low

3-5. (D) Radar to radar interference

Lesson 3. Points to Ponder

1. Note the radar Range. Note the display mode. Look for prominent radar targets on the screen. Note their approximate range and relative bearings. Look out the window to find them. If you can't or it is night, look to the chart. Need to have your DR track plotted on the chart for orientation. Then start with numerical radar ranges and bearings to be more quantitative. [With echarts in use, set range rings on the vessel icon to match the radar screen. See what they intersect (bearings and ranges) then see them on the radar screen. The position projection vector on the echart is your heading line.]

2. HBW smears out all targets half the HBW to the right and left. Thus buoys get made larger. Tangent bearings will be off by HBW/2. Any two features closer than HBW will be smeared together. Thus a distant channel might not show, or two close islets would appear as one. A tug and tow might show as one large target.

3. Minimum azimuthal width is HBW. Minimum radial depth is about equal to pulse length. (Theoretical limit is half the pulse length, but in practice depth of small targets will be at least equal to the pulse length.)

4. Horizontal shadows block seeing around the corner of any tall landmass, thus artificially shortening a target in the azimuthal (bearing) direction. Vertical shadows block seeing lower features behind taller ones, thus shortening the depth of a target in the radial direction.

5. The effect of a vertical and horizontal shadow of a tall protrusion of the shoreline. The lack of target (the "water") on the inshore side of the "ship" is caused by the vertical shadow. The two ends of the ship are caused by the horizontal shadows. When you get a bit farther along the shore the target shape will change.

6. Versatile ECS displays are a boon to radar interpretation. The position projector line on the vessel icon is your heading line on the radar. Set range rings on the vessel icon and you are carrying along your radar screen right on the chart as your vessel icon moves along. You see in an instant what should be visible at what ring and what bearing.

7. Side lobe interference from close targets smears the target radially around the screen. Reduce Gain to fix it. Radar to radar interference in harbors or in busy channels shows up as transient spirals or fields of dashes. The IR control will remove these completely. (Any other likely interference would require some large structure on your vessel in line with the antenna and the presence of a large target.)

Lesson 4. Position Navigation

4-1. (C) A RACON signal

4-2. (B) Fast but not as accurate as other methods

4-3. (C) Intersecting ranges from 2 or more targets

4-4. (A) Range and bearing to a single target

4-5. (D) Setting range rings on the vessel icon to match those on the radar

4-6. (A) 25 to 30 degrees, more or less independent of antenna size

Lesson 4. Points to Ponder

1. Quickest is to read range and bearing from ECS display or paper chart to prominent radar target, then turn to radar to confirm that observation.

2. Two crossed bearings: least precise. Range and bearing to single target: quick, better than bearings alone. Range and two tangents, better still. Two or more intersecting ranges; best possible radar fix.

3. Bearings susceptible to error from: compass or heading-sensor errors, antenna-alignment errors, spreading of the target by HBW, and the challenge of recording the heading at the exact time of the bearing measurement. Also the heeling angle of a sailing vessel can distort the bearing in some directions.

4. Three range fix is most accurate. Use VRM or cursor to get ranges to nearest edges of 3 prominent landmarks, then plot those ranges as circles of position from the target points. The intersection is the fix. (When moving, measure beam targets first as their ranges change the most slowly.)

5. When GPS is not available, as in narrow steep channels or fjords; when there is no adequate chart for the location, as when anchoring in a small cove; when there is doubt about chart datum matches between GPS and charts in use; when the radar fix is adequate and efforts of Lat/Lon plotting are not justified.

6. More than a few yards away from a typical small craft radar with the antenna rotating would likely be considered safe by most safety standards for transient exposure. Up close to the antenna, on the other hand, looking into it at eye level, would definitely cause an exposure higher than recommended standards. (There are formulas and standards presented in the text for numerical estimates based on specific radar systems.) Simple prudence calls for avoiding exposure to the radar beam whenever possible. A person might chose to minimize all exposure, pointing out that published safe levels for radiation exposure *at all frequencies* have a tendency to go down with time.

Lesson 5. Radar Piloting

5-1. (C) VRM

5-2. (B) EBL

5-3. (A) Heading line

5-4. (D) ERBL

5-5. (C) VRM

5-6. (D) ERBL

Lesson 5. Points to Ponder

1. Geographic range = 1.2 x sqrt (antenna height) + 1.2 x sqrt (target height). Heights in feet, range in nmi. This is maximum range, we actually need to be a bit closer so we get enough of the target over the horizon.

2. Maximum range is determined by power output and is a fixed specification of the unit; geographic range is limited by antenna height and target height and size.

3. The simplest way is to just set the VRM at the safe distance you want to maintain and then round without letting the VRM go ashore.

4. To maintain a route parallel to a shoreline, or to insure that you are pointed toward your proper destination. It is also a way to check that you are staying on course and not being set by a current.

5. You can find a natural range by aligning heading line or EBL with the edges or centers of two radar targets. This can give you an LOP, which can also be used to watch that you are not being set by a current. You can also use the EBL to monitor a danger bearing in the fog once it is known.

6. Gives you an overview of a cove that might not show details on the chart. Shows locations rocks awash, logs, or other anchored vessels. Shows when you are in the middle of the cove, or adequately far from another vessel or the shoreline. You can also check to see if you are dragging anchor. An alarm circle set around you will warn of any approaching vessels at night, or if you have dragged closer to the shore.

7. You are getting set by current in a direction away from the buoy.

Lesson 6. Evaluating risk of collision

Detailed solutions and plots on page 36

6-1 A. (a) 1.9 nmi, 1318; (b) 2.5 kt, 130 M; (c) W

6-2 B. (a) 1.2 nmi, 1306; (b) 9.0 kt, 310 M; (c) G

6-3 C. (a) 0.8 nmi, 1330; (b) 14.0 kt, 130 M; (c) G

6-4 D. (a) 1.0 nmi, 1305; (b) 9.3 kt, 220 M; (c) R&G

6-5 E. (a) 1.5 nmi, 1300; (b) 8.0 kt, 130 M; (c) W

6-6 F. (a) 2.2 nmi, 1309; (b) 0.0 kt, ------ ; (c) ?

6-7 G. (a) 1.5 nmi, 1021; (b) 13.0 kt, 093 M; (c) R&G

6-8 H. (a) 1.5 nmi, 1056; (b) 7.7 kt, 028 M; (c) W

6-9 I. (a) 0.7 nmi, 1030; (b) 13 kt, 019 M; (c) R

6-10 J. (a) 0.9 nmi, 1054; (b) 7.5 kt, 069 M; (c) W

6-11 K. (a) 0.5 nmi, 1052; (b) 10.5 kt, 063 M; (c) G

6-12 L. (a) 2.2 nmi, 1022; (b) 15.5 kt, 071 M; (c) R

6-13 M. (a) 0.8 nmi, 1447; (b) 12.3 kt, 201 T; (c) R

6-14 N. (a) 0.0 nmi, 1513; (b) 9.0 kt, 217 T; (c) W

6-15 O. (a) 1.5 nmi, 1427; (b) 2.4 kt, 050 T; (c) R

ANSWERS

Lesson 6. Points to Ponder

1. Set EBL and VRM on the target. Turn on the plot (wake) option. Note this would be the procedure regardless of where the target first appears on the screen. Then if it is night, start to look for it with binoculars. In a crucial situation, it would also be good to mark the position on the screen by hand as well.

2. When first seen far off we do not have enough of a wake history to project all the way to the center for a good CPA. Also in Head-up display, the target position will be smeared out as we yaw about. The farther the target, the larger these smears. It must get closer before we can know enough to maneuver. We need two well defined positions (range, bearing, and time) to identify the DRM and CPA of the target.

3. That the course and speed of both the target and ourselves remains constant. We also assume that any significant current that we and the target are experiencing remains the same as well.

4. When a target is stopped, it moves straight down the radar screen, just as buoy does, regardless of our speed. Thus when the target vessel slows down, it moves more like it is stopped, i.e. the target trail will turn (to some extent) down screen. The opposite occurs when the target vessel speeds up.

5. It curves up screen. Recall that the true direction of a target is always aft of its apparent direction, i.e. more up-screen. When we stop it shifts completely to its true direction, so when we slow down it moves toward that direction, which is up-screen. The opposite occurs when we speed up.

6. The buoy trail is a measure of our speed. The down-screen trail has SRM of twice our speed, but half of that is due to our motion. Thus the down-screen target is a vessel of our speed moving opposite to our course. The up-screen target must be headed in the same direction we are. And he, like the other one, is making a relative speed of twice our speed, but we are still falling behind, so his true speed is 3 times our speed.

7. Draw a line straight down screen from the tail of the wake with a length equal to one tenth of your speed. Connect the bottom of the line to the present target position. The direction of that line (projected beyond the target position) is the true course of the target. The length of that line is the true speed of the target. (In North-up mode, you get its true course directly; in Head-up mode you measured a relative course. You must add your compass heading to obtain the target's compass course)

8. (1) Main concern is that our knotmeter and heading sensor remain properly calibrated. (2) If we lose the target and then re-aquire it, we must be certain that it is the same target... always a concern if two targets pass close by.

Lesson 7. Radar and the Navigation Rules

7-1. (C) 60°

7-2. (B) In the fog we do not always hold course and speed when being overtaken

7-3. (D) There is no right of way in the fog, for power or for sail

7-4. (A) Turn right

7-5. (A) Turn right

7-6. (A) Turn right

Lesson 7. Points to Ponder

1. A proper look out just has to know enough about radar to identify a converging target; whereas the person judging risk of collision has to know how to evaluate that risk and maneuver according to the Rules.

2. (i) general limitations, i.e. basic tuning procedures must be understood, (ii) range, i.e. might miss approaching targets if looking only at low ranges, (iii) sea state or rain can hide small targets, (iv) small targets may not be seen on large ranges, (v) difficulties of tracking and evaluating multiple targets, (vi) use radar to determine the visible range as needed.

3. Knowing these limitations on the radar watch of a ship spotting our small vessel, we cannot rely on them detecting us. If there is any doubt about the interaction, with your radar, find the distance and bearing to the ship (which you will see easily) and call them on the VHF to report your position (as range and bearing), course and speed. And in any event, navigate with due caution and obey the rules on maneuvering and lights, so as not to confuse them. Shining bright lights on your sails is perfectly legal.

4. It means at minimum set the EBL and VRM on the target and start the plot option (or mark the target position and label it with the time). Next best is to obtain at least an estimated solution of the relative motion diagram to determine the target's actual course and speed.

5. Mostly this means do not maneuver until you know for sure what the other vessel is doing. Which in turn means accumulate enough of a wake record (or hand-marked trail history) to be able to get a dependable CPA. At a long distance off this measurement is often not reliable, especially in Head-up mode.

(It is quite possible to see a smear of a wake coming right down the heading line a long way off in head-up mode. To then prematurely turn right might in fact put you right in line with a vessel that was in fact going to pass safely on your starboard side.)

Also remember that visual information is almost always a great asset to interpreting the radar, just as the reverse is true as well. We do not want to get stuck with our head in the radar screen if we could have done better without radar at all! It is not uncommon for someone at the radar screen to become very nervous about an interaction, whereas everyone else on deck is simply watching a ship go by.

6. Turn right for converging targets forward of the beam; turn away for those aft of the beam. (On the starboard beam consider significant speed reduction as an alternative.)

7. It is the space around my vessel in the present conditions that I need to be able to maneuver to avoid a collision, regardless of what the other vessel might (could) do, suddenly and unexpectedly.

8. Know and obey the Navigation Rules — because virtually every collision involves the violation of at least one rule on the part of both vessels in the collision.

Radar Abbreviations	
AC	Anti Clutter
AIS	Automatic Identification System
ARPA	Automatic Radar Plotting Aid
ATA	Automatic Tracking Aid
CPA	Closest Point of Approach
C-U	Course-up display mode
DRM	Direction of Relative Motion
DRM	Dead Reckoning
EBL	Electronic Bearing Line
ECS	Electronic Charting System
ERBL	Electronic Range and Bearing Line
ES	Echo Stretch
FTC	Fast Time Constant = AC Rain
GPS	Global Positioning System
HBW	Horizontal Beam Width
H-U	Head-up display mode
IR	Interference Rejection
LOP	Line of Position
N-U	North-up display mode
R	Range or Relative bearing
RACON	RAdar BeaCON
RFM	*Radar for Mariners* text
RR	Range Rings
SART	Search and Rescue Radar Transponder
SRM	Speed of Relative Motion
STC	Sensitivity Time Control = AC Sea
TCPA	Time to CPA
VRM	Variable Range Marker
RMD	Relative Motion Diagram

ANSWERS

Lesson 6. Evaluating risk of collision

Detailed solutions and plots

6-1 A. (a) 1.9 nmi, 1318;   (b) 2.5 kt, 130 M;   (c) W

Plot target at appropriate ranges and bearings; note times. Draw DRM. Draw CPA perpendicular to the extended DRM and read CPA range as 1.9 nm. Calculate SRM from distance between 1254 and 1300 contacts, as .55nm in 6 min = 5.5 kts SRM. Calculate time to CPA as 1.7nm at 5.5 kts = 18 minutes. Calculate TCPA as 1300 + 18 = 1318 TCPA. Observe that downscreen target Contact A indicates a distance between its 1254 and 1300 contacts that is less than buoy trail; therefore, A must be a vessel proceeding on same course but at a speed less than own ship; that is, we are overtaking Contact A. Contact A's speed is the difference between 6-minute buoy trail and distance between 6-minute contacts, as .8nm minus .55nm = .25nm, or converting from 6-minutes, 2.5 kts. As an overtaking vessel we would view Contact A's stern light (W).

6-2 B. (a) 1.2 nmi, 1306;   (b) 9.0 kt, 310 M;   (c) G

Plot target at appropriate ranges and bearings; note times. Draw DRM. Draw CPA perpendicular to the extended DRM and read CPA range as 1.2 nm. Calculate SRM from distance between 1254 and 1300 contacts, as 1.7nm in 6 min = 17 kts SRM. Calculate time to CPA as 1.7nm at 17 kts = 6 minutes. Calculate TCPA as 1300 + 6 = 1306 TCPA. Observe that Contact B is a downscreen target, also that the distance between the 1254 and 1300 contacts is greater than buoy trail; therefore, Contact B must be proceeding on course opposite to own ship. Contact B's speed is the difference between 6-minute buoy trail and distance between 6-minute contacts, as .1.7nm minus .8nm = .9nm, or 9 kts. We view opposite direction traffic's starboard sidelight (G).

6-3 C. (a) 0.8 nmi, 1330;   (b) 14.0 kt, 130 M;   (c) G

Plot target at appropriate ranges and bearings; note times. Draw DRM. Draw CPA perpendicular to the extended DRM and read CPA range as .8 nmi. Calculate SRM from distance between 1254 and 1300 contacts, as .63 nmi in 6 min = 6.3 kts SRM. Calculate time to CPA as 3.1 nmi at 6.3 kts = 30 minutes. Calculate TCPA as 1300 + 30 = 1330 TCPA. Observe that Contact C is an upscreen target, clearly overtaking own ship. Contact C's course is same as own ship, and speed is the sum of 6 min buoy trail + 6-min contact trail, converted to kts, as .8 + .63 = 1.4 = 14 kts. We view this contact's starboard sidelight (G).

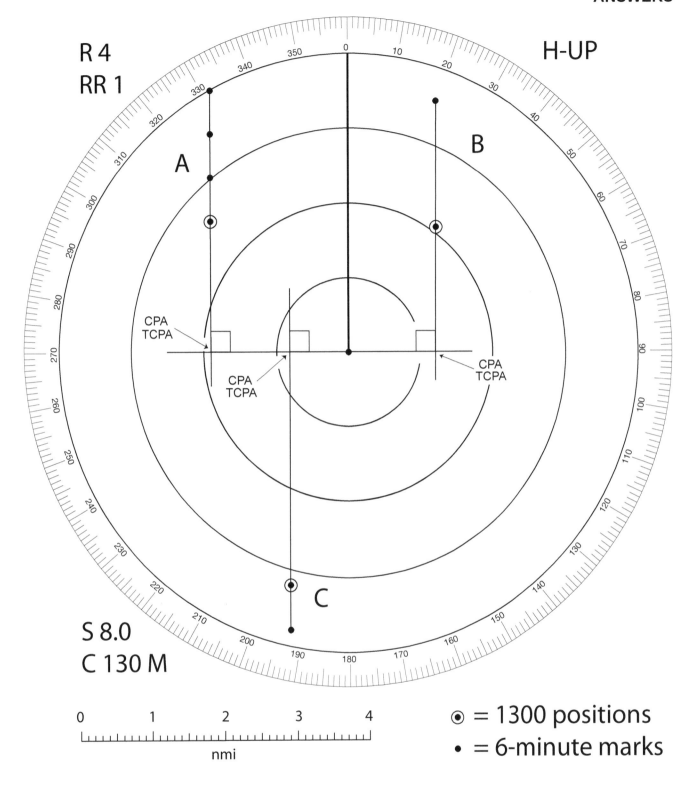

R 4
RR 1

H-UP

A

B

CPA
TCPA

CPA
TCPA

CPA
TCPA

S 8.0
C 130 M

C

0   1   2   3   4
nmi

⊙ = 1300 positions
• = 6-minute marks

ANSWERS

6-4 D. (a) 1.0 nmi, 1305;   (b) 9.3 kt, 220 M;   (c) R&G

Plot target at appropriate ranges and bearings; note times. Draw DRM. Draw CPA perpendicular to the extended DRM and read CPA range as 1.0 nmi. Calculate SRM from distance between 1254 and 1300 contacts, as 1.23 nmi in 6 min = 12.3 kts SRM. Calculate time to CPA as 1 nmi at 12.3 kts = 5 minutes. Calculate TCPA as 1300 + 5 = 1305 TCPA. Draw 6-minute buoy trail astern from Contact D's 1248 plot. Complete the RMD and read target's actual course as 090 R; convert to Magnetic as 130 + 090 = 220 M. Read target's distance over 6-minutes as .93 nmi; convert to 9.3 kts. As at 1300, our view of this 090 R target is bow-on; that is, red and green sidelights (R & G).

6-5 E. (a) 1.5 nmi, 1300;   (b) 8.0 kt, 130 M;   (c) W

Plot target at appropriate ranges and bearings; note times. Observe that Contact E is a 'buddy boat' traveling at same course and speed. Therefore, it's present range is its CPA. Our position off the stern quarter of Contact E would offer us a stern light (W).

6-6 F. (a) 2.2 nmi, 1309;   (b) 0.0 kt, ------ ;   (c) ?

Plot target at appropriate ranges and bearings; note times. Draw DRM. Draw CPA perpendicular to the extended DRM and read CPA range as 2.2 nmi. Calculate SRM from distance between 1254 and 1300 contacts, as 0.8 nmi in 6 min = 8 kts SRM. Calculate time to CPA as 1.2 nmi at 8 kts = 9 minutes. Calculate TCPA as 1300 + 9 = 1309 TCPA. Observe that Contact SRM = buoy trail; therefore, target is a buoy or is a vessel underway but not making way. Refer to chart for buoy lighting. If a vessel, aspect and lighting are undetermined.

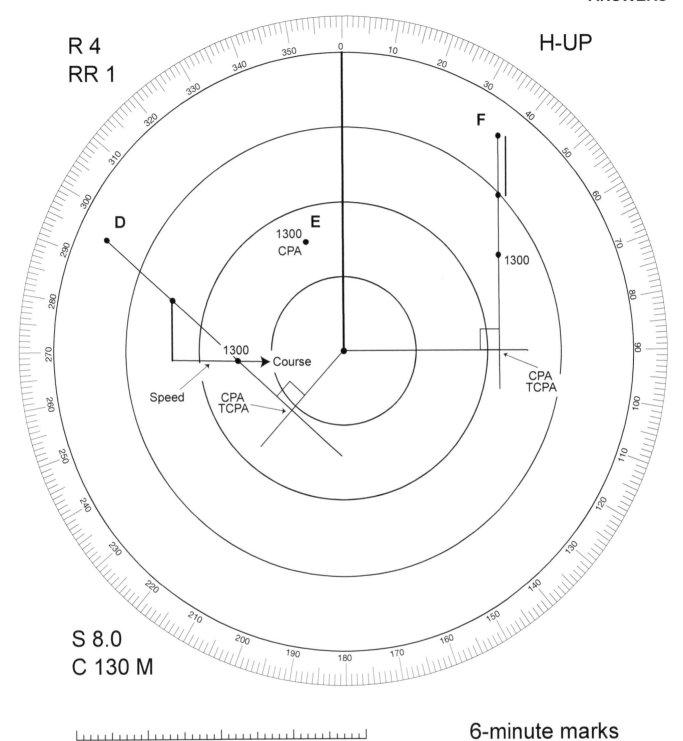

R 4
RR 1

H-UP

F

D

E

1300 •
CPA

1300

1300 •
Course

Speed

CPA
TCPA

CPA
TCPA

S 8.0
C 130 M

6-minute marks

ANSWERS

6-7 G. (a) 1.5 nmi, 1021;   (b) 13.0 kt, 093 M;   (c) R&G

Plot target at appropriate ranges and bearings; note times. Draw DRM. Draw CPA perpendicular to the extended DRM and read CPA range as 1.5 nmi. Calculate SRM from distance between 1006 and 1012 contacts, as 1 nmi in 6 min = 10 kts SRM. Calculate time to CPA as 1.55 nmi at 10 kts = 9 minutes. Calculate TCPA as 1012 + 9 = 1021 TCPA. Draw 6-minute buoy trail astern from Contact G's 1006 plot. Complete the RMD and read target's actual course as 048 Relative; convert to Magnetic as 045 + 048 = 093 M. Read target's distance over 6-minutes as 1.3 nmi; convert to 13 kts. As at 1300, our view of this target is bow-on; that is, red and green sidelights (R & G).

6-8 H. (a) 1.5 nmi, 1056;   (b) 7.7 kt, 028 M;   (c) W

Plot target at appropriate ranges and bearings; note times. Draw DRM. Draw CPA perpendicular to the extended DRM and read CPA range as 1.5 nmi. Calculate SRM from distance between 1006 and 1012 contacts, as .3 nmi in 6 min = 3 kts SRM. Calculate time to CPA as 2.2 nmi at 3 kts = 44 minutes. Calculate TCPA as 1012 + 44 = 1056 TCPA. Draw 6-minute buoy trail astern from Contact H's 0948 plot. Complete the RMD and read target's actual course as 343 Relative; convert to Magnetic as (045 + 343) - 360 = 028 M. Read target's distance over 6-minutes as .77 nmi; convert to 7.7 kts. We calculate that as at 1012, own ship bears 146 R from target; thus we view of this target's stern light (W).

6-9 I. (a) 0.7 nmi, 1030;   (b) 13 kt, 019 M;   (c) R

Plot target at appropriate ranges and bearings; note times. Draw DRM. Draw CPA perpendicular to the extended DRM and read CPA range as .7 nmi. Calculate SRM from distance between 1006 and 1012 contacts, as .62 nmi in 6 min = 6.2 kts SRM. Calculate time to CPA as 1.8 nmi at 6.2 kts = 18 minutes. Calculate TCPA as 1012 + 18 = 1030 TCPA. Draw 6-minute buoy trail astern from Contact I's 1000 plot. Complete the RMD and read target's actual course as 334 R; convert to Magnetic as (045 + 334) - 360 = 019 M. Read target's distance over 6-minutes as 1.3 nmi; convert to 13 kts. As at 1012 we view this contact's port sidelight (R).

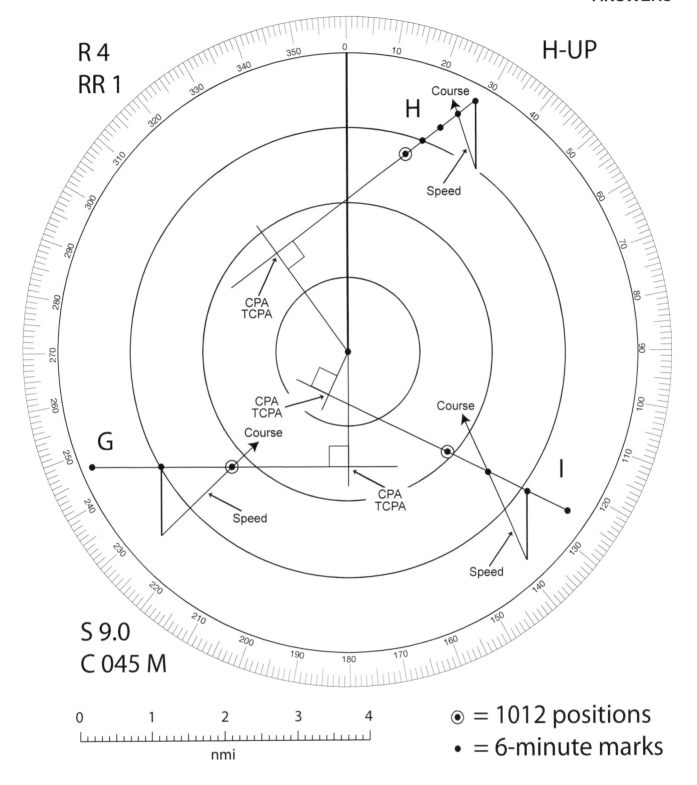

R 4
RR 1

H-UP

H

Course

Speed

CPA
TCPA

CPA
TCPA

G

Course

CPA
TCPA

Course

I

Speed

Speed

S 9.0
C 045 M

0   1   2   3   4
nmi

⊙ = 1012 positions
• = 6-minute marks

ANSWERS

6-10 J. (a) 0.9 nmi, 1054;   (b) 7.5 kt, 069 M;   (c) W

Plot target at appropriate ranges and bearings; note times. Draw DRM. Draw CPA perpendicular to the extended DRM and read CPA range as .9 nmi. Calculate SRM from distance between 1006 and 1012 contacts, as .74 nmi in 12 min = 3.7 kts SRM. Calculate time to CPA as 2.6 nmi at 3.7 kts = 42 minutes. Calculate TCPA as 1012 + 42 = 1054 TCPA. Draw 12-minute buoy trail astern from Contact J's 1006 plot. Complete the RMD and read target's actual course as 024 R; convert to Magnetic as 045 + 024 = 069 M. Read target's distance over 12-minutes as 1.5 nmi; convert to 7.5 kts. Contact J's relative bearing to own ship is 145 R, less target's relative heading 024 = 121 R off Contact J's bow, well aft of the 112.5 degree line; therefore, we view Contact J's stern light (W).

6-11 K. (a) 0.5 nmi, 1052;   (b) 10.5 kt, 063 M;   (c) G

Plot target at appropriate ranges and bearings; note times. Draw DRM. Draw CPA perpendicular to the extended DRM and read CPA range as .5 nmi. Calculate SRM from distance between 1006 and 1012 contacts, as .66 nmi in 12 min = 3.3 kts SRM. Calculate time to CPA as 2.2 nmi at 3.3 kts = 40 minutes. Calculate TCPA as 1012 + 40 = 1052 TCPA. Draw 12-minute buoy trail astern from Contact K's 1006 plot. Complete the RMD and read target's actual course as 018 R; convert to Magnetic as 045 + 018 = 063 M. Read target's distance over 12-minutes as 2.1 nmi; convert to 10.5 kts. We view Contact K as a starboard sidelight (G).

6-12 L. (a) 2.2 nmi, 1022;   (b) 15.5 kt, 071 M;   (c) R

Plot target at appropriate ranges and bearings; note times. Draw DRM. Draw CPA perpendicular to the extended DRM and read CPA range as 2.2 nmi. Calculate SRM from distance between 1006 and 1012 contacts, as 1.7 nmi in 12 min = 8.5 kts SRM. Calculate time to CPA as 1.4 nmi at 8.5 kts = 10 minutes. Calculate TCPA as 1012 + 10 = 1022 TCPA. Draw 12-minute buoy trail astern from Contact L's 1006 plot. Complete the RMD and read target's actual course as 026 R; convert to Magnetic as 045 + 026 = 071 M. Read target's distance over 12-minutes as 3.1 nmi; convert to 15.5 kts. We view Contact L as a port sidelight (R).

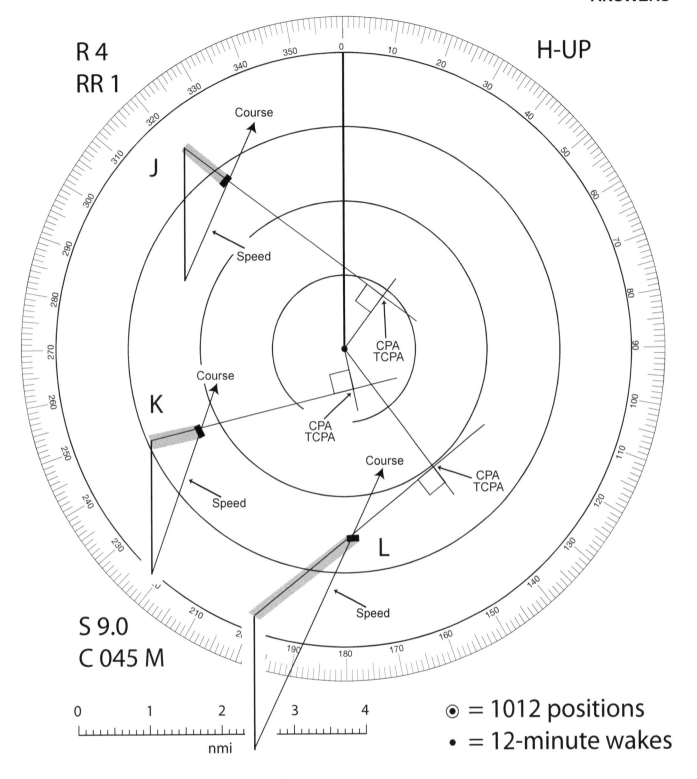

R 4
RR 1

H-UP

J

Course

Speed

CPA
TCPA

CPA
TCPA

K

Course

Speed

Course

CPA
TCPA

L

Speed

S 9.0
C 045 M

0   1   2   3   4
nmi

⊙ = 1012 positions
• = 12-minute wakes

ANSWERS

6-13 M. (a) 0.8 nmi, 1447;   (b) 12.3 kt, 201 T;   (c) R

Plot target at appropriate ranges and bearings; note times. Draw DRM. Draw CPA perpendicular to the extended DRM and read CPA range as 0.8 nmi. Calculate SRM from distance between 1414 and 1420 contacts, as 0.6 nmi in 6 min = 6.0 kts SRM. Calculate time to CPA as 2.7 nmi at 6 kts = 27 minutes. Calculate TCPA as 1420 + 27 = 1447 TCPA. Draw 6-minute buoy trail astern from Contact M's 1414 plot. Complete the RMD and read target's actual course as 201 T. Read target's distance over 6-minutes as 1.23 nmi; convert to 12.3 kts. We view Contact M's port sidelight (R).

6-14 N. (a) 0.0 nmi, 1513;   (b) 9.0 kt, 217 T;   (c) W

Plot target at appropriate ranges and bearings; note times. Draw DRM. Draw CPA perpendicular to the extended DRM and read CPA range as nil; collision risk exists. Calculate SRM from distance between 1414 and 1420 contacts, as .37 nmi in 6 min = 3.7 kts SRM. Calculate time to CPA as 3.3 nmi at 3.7 kts = 53 minutes. Calculate TCPA as 1420 + 53 = 1513 TCPA. Draw 6-minute buoy trail astern from Contact N's 1414 plot. Complete the RMD and read target's actual course as 217 T. Read target's distance over 6-minutes as .9 nmi; convert to 9 kts. We view Contact N's stern light (W).

6-15 O. (a) 1.5 nmi, 1427;   (b) 2.4 kt, 050 T;   (c) R

Plot target at appropriate ranges and bearings; note times. Draw DRM. Draw CPA perpendicular to the extended DRM and read CPA range as 1.5 nmi. Calculate SRM from distance between 1414 and 1420 contacts, as 1.44 nmi in 6 min = 14.4 kts SRM. Calculate time to CPA as 1.7 nmi at 14.4 kts = 7 minutes. Calculate TCPA as 1420 + 7 = 1427 TCPA. Draw 6-minute buoy trail astern from Contact O's 1414 plot, and observe that Contact's direction is either same as or reciprocal of own ship. Observe that contact's relative distance closing over 6 minutes is greater than buoy trail; therefore, target is opposite direction traffic at 050 T. Calculate Contact O's speed as the difference between its 6-minute distance and buoy trail, as 1.44 nmi - 1.2 nmi = .24 nmi; convert to 2.4 kts. We view Contact O's port sidelight (R).

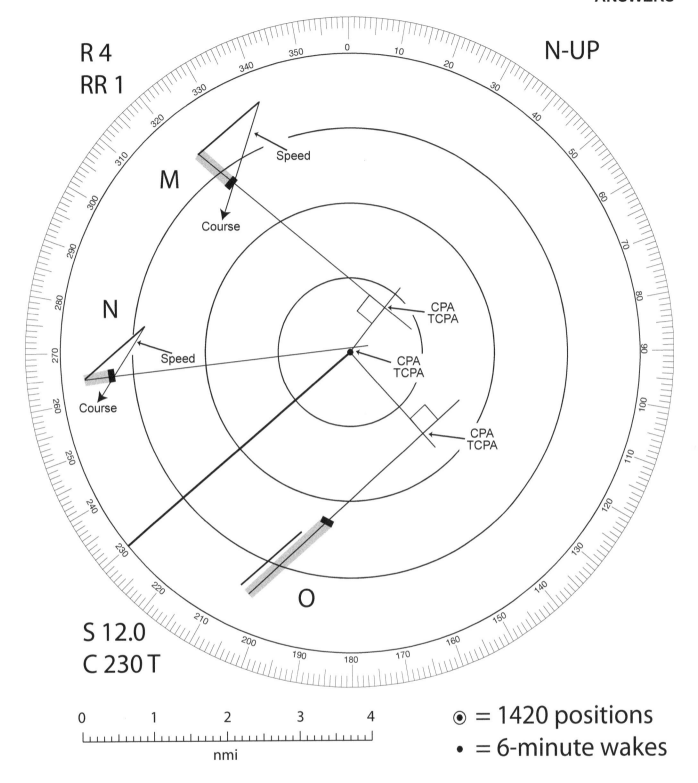

R 4
RR 1

N-UP

M

N

Speed

Course

Speed

Course

CPA
TCPA

CPA
TCPA

CPA
TCPA

O

S 12.0
C 230 T

0   1   2   3   4
|⌊⌊⌊⌊⌊|⌊⌊⌊⌊⌊|⌊⌊⌊⌊⌊|⌊⌊⌊⌊⌊|
nmi

⊙ = 1420 positions
• = 6-minute wakes

RADAR PLOTTING SHEET

The sheet may be duplicated for practice with radar plotting. (5-ring sample on page 104)

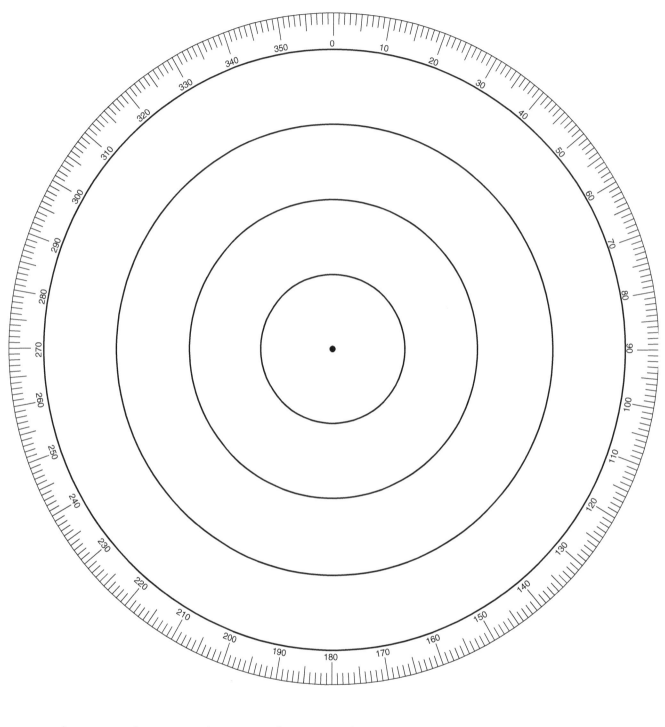

Advanced Problems in Radar Maneuvering

Background

Advanced Problems in Radar Maneuvering are presented here for the professional mariner and the dedicated recreational skipper. A mere glance through these problems will give the reader a taste of how complex multiple ship maneuvering can be, with speed changes, own ship maneuvering, target maneuvering, and other actions contributing to the age old challenge of keeping two vessels from attempting to occupy the same bit of water at the same time. At a minimum, we hope the reader will take away from a review of this advanced section a commitment to greater vigilance on watch and a dedicated adherence to the Navigation Rules.

Methodology

The problems were first solved using the graphics plotting capability of an electronic charting system, in this case Coastal Explorer© (CE); from those solutions the publication graphics and narratives were created. Other typical charting software could have been used equally well.

CE offers more or less precision depending on how 'deep' one zooms into the electronic chart to place the initial target locations. Working the Relative Motion Diagram on a 1:100,000 scale chart may appear accurate, but when one zooms in slight misalignments are apparent. For this publication we took care to use 1:10,000 scale or better for the sake of accuracy; however at sea such care likely would have added unnecessary delay in analyzing a dynamic collision avoidance situation.

These advanced maneuvering problems are presented in order of increasing complexity. Some involve situations that take place over a large area, perhaps diagrammed on a plotting sheet over 20 miles range, yet the result appears on the page in front of you barely 6-inches across. Such are the limitations of publication; nevertheless we trust that these graphics together with a narrative of the solution will be helpful to the marine radar student.

Our narratives of the solutions become progressively simpler as the problems become more difficult. This is to avoid redundancy in describing steps that have been well explained earlier in the text, tasks such as measuring and computing SRM and TCPA. These basic steps are described in detail in Chapter 6, but later are referred to only generally.

Courses and bearings are assumed to be referenced to True, unless otherwise stated. N-UP mode was used throughout. Other display modes such as H-UP could have been used, but the resulting graphic of the solution would have been overly complicated. The advanced problems are excerpted from Publication 1310, but the solutions presented here are our own.

1. Own ship, on course 311°, speed 17 knots, obtains the following radar bearings and ranges at the times indicated, using a radar setting of 24 miles:

Time	Bearing	Range (mi.)
1136	280°	16.0
1142	274°	13.6
1148	265°	11.4

Required—

(a) Range at CPA

(b) Time at CPA

(c) Direction of Relative Motion (DRM)

Answers. (a) 8.2 nmi, (b) 1204.5, (c) 131

Plot target at appropriate ranges and bearings; note times. Draw DRM line and read DRM (c). Draw CPA line perpendicular to the extended DRM, and read range at CPA (a). Measure distance between 1136 and 1148 contacts over 12 min and calculate SRM; then calculate time to CPA and TCPA (b).

Radar Maneuvering Problem 1.

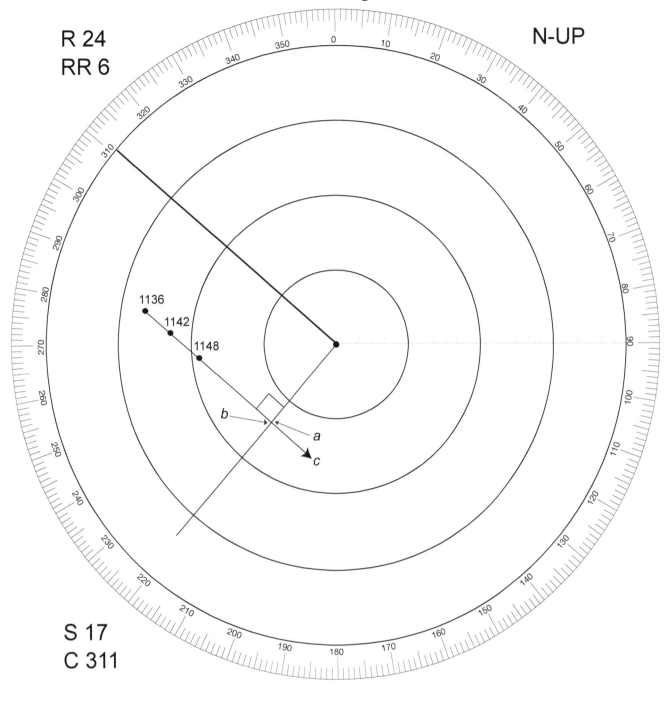

R 24
RR 6

N-UP

S 17
C 311

49

2. Own ship, on course 000°, speed 12 knots, obtains the following radar bearings and ranges at the times indicated, using a radar range setting of 12 miles:

Time	Bearing	Range (mi.)
0410	035°	11.1
0416	031°	9.2
0422	025°	7.3

Required—

(a) Distance at which the contact will cross dead ahead.

(b) Direction of relative movement (DRM).

(c) Speed of relative movement (SRM); relative speed.

(d) Range at CPA.

(e) Bearing of contact at CPA.

(f) Relative distance (MRM) from 0422 position of contact to the CPA.

(g) Time at CPA.

(h) Distance own ship travels from the time of the first plot (0410) to the time of the last plot (0422) of the contact.

(i) True course of the contact.

(j) Actual distance traveled by the contact between 0410 and 0422.

(k) True speed of the contact.

Answers. Assuming that the contact maintains course and speed: (a) 4.3 nmi, (b) 233, (c) 20.6 kt, (d) 3.4 nmi, (e) 323, (f) 6.4 nmi, (g) 0441, (h) 2.4 nmi, (i) C 269, (j) 3.3 nmi, (k) 16 kt

Plot target at appropriate ranges and bearings; note times. Draw DRM line. Read range that target will cross dead ahead (a). Read DRM (b). Calculate SRM (c) from distance between 0410 and 0422 contacts. Draw CPA line perpendicular to the extended DRM and read range (d). Read bearing at CPA (e). Measure distance from 0422 contact to CPA (f). Calculate TCPA (g). Calculate distance own ship travels from first target plot to most recent target plot (h). Draw 12-minute buoy trail astern from first target plot. Complete the RMD and read true course of contact (i). Measure actual distance traveled by contact between 0410 and 0422 (j). Calculate true speed of contact (k).

Radar Maneuvering Problem 2.

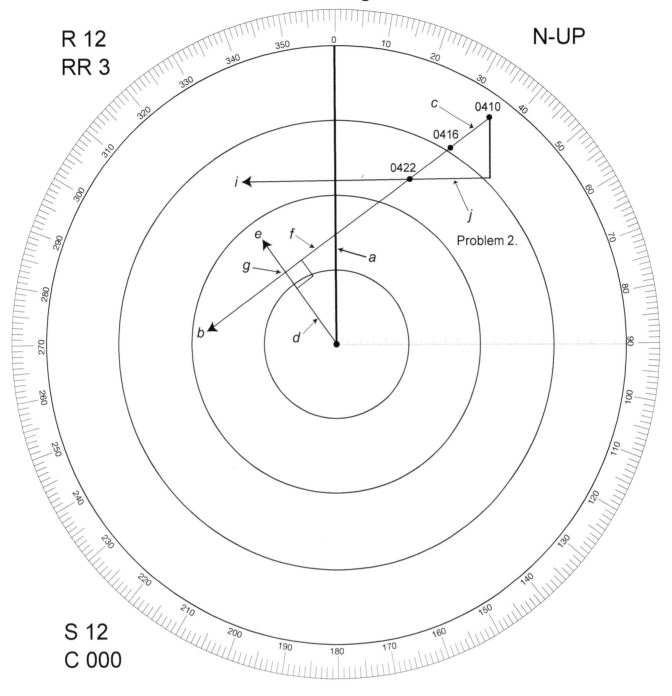

3. Own ship, on course 030°, speed 23 knots, obtains the following radar bearings and ranges at the times indicated, using a radar range setting of 12 miles:

Time	Bearing	Range (mi.)
1020	081°	10.8
1023	082°	9.2
1026	083°	7.7

Required—

(a) Range at CPA.

(b) Bearing of contact at CPA.

(c) Speed of relative movement (SRM); relative speed.

(d) Time at CPA.

(e) Distance own ship travels from the time of the first plot (1020) to the time of the 1026 plot of the contact; distance own ship travels in 6 minutes.

(f) True course of the contact.

(g) Actual distance traveled by the contact between 1020 and 1026.

(h) True speed of the contact.

(i) Assuming that the contact has turned on its running lights during daylight hours because of inclement weather, what side light(s) might be seen at CPA?

Answers. Assuming that the contact maintains course and speed: (a) .9 nmi, (b) 166, (c) 31 kt, (d) 1041, (e) 2.3 nmi, (f) C 304, (g) 2.3 nmi, (h) 23 kt, (i) Starboard (green) side light

Plot target at appropriate ranges and bearings; note times. Draw DRM line. Draw line perpendicular to the extended DRM and read range at CPA (a). Read bearing at CPA (b). Measure distance between 1020 and 1026 contacts and calculate SRM (c). Measure distance to CPA and calculate TCPA (d). Calculate distance own ship travels over 6 minutes (e). Draw 6-minute buoy trail astern from first contact plot. Complete the RMD and read true course of contact (f). Measure actual distance contact traveled over 6 minutes (g). Calculate actual speed of contact (h). Determine aspect of target at CPA and lights (i).

Radar Maneuvering Problem 3.

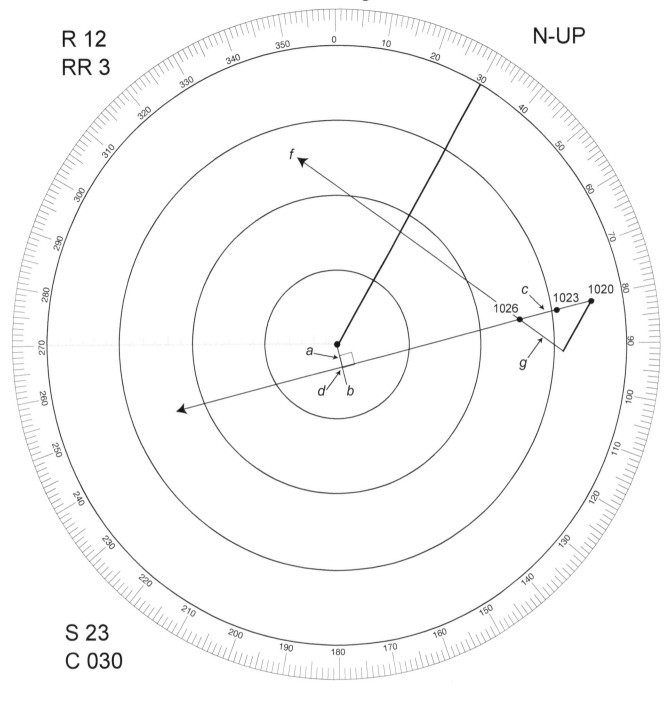

R 12
RR 3

N-UP

S 23
C 030

4. Own ship, on course 000°, speed 11 knots, obtains the following radar bearings and ranges at the times indicated, using a radar range setting of 12 miles:

Time	Bearing	Range (mi.)
1100	080°	12.0
1106	080°	10.8
1112	080°	9.6

Required—

(a) Range at CPA.

(b) Speed of relative movement (SRM); relative speed.

(c) Time at CPA.

(d) True course of contact.

Decision.—When the range to the contact decreases to 6 miles, own ship will change course so that the contact will pass safely ahead with a CPA of 2.0 miles.

Required—

(e) New course for own ship.

(f) New SRM after course change.

Answers. Assuming that the contact maintains course and speed: (a) Nil; risk of collision exists, (b) 12 kt, (c) 1200, (d) C 307, (e) C 061, (f) 22 kt

Plot target at appropriate ranges and bearings; note times. Draw DRM line. Read range at CPA (a), which is nil; collision risk exists. Measure distance between first and last contacts; calculate SRM (b). Measure distance to CPA; calculate TCPA (c). Draw 12-minute buoy trail astern from first target plot. Complete the RMD and read true course of contact (d). Draw 2 nmi desired CPA range ring. Draw new DRM proceeding from target 6 nmi range to pass ahead of own ship and tangent to the 2 nmi CPA ring. Parallel the new DRM to intersect the most recent target plot. Draw an arc of the buoy trail to intersect the new DRM line; read new course for own ship (e). Note that the target's actual course and speed line, together with the new DRM line and the arc of the buoy trail complete a new RMD. Measure distance from (e) to 1112 plot; calculate new SRM (f).

Radar Maneuvering Problem 4.

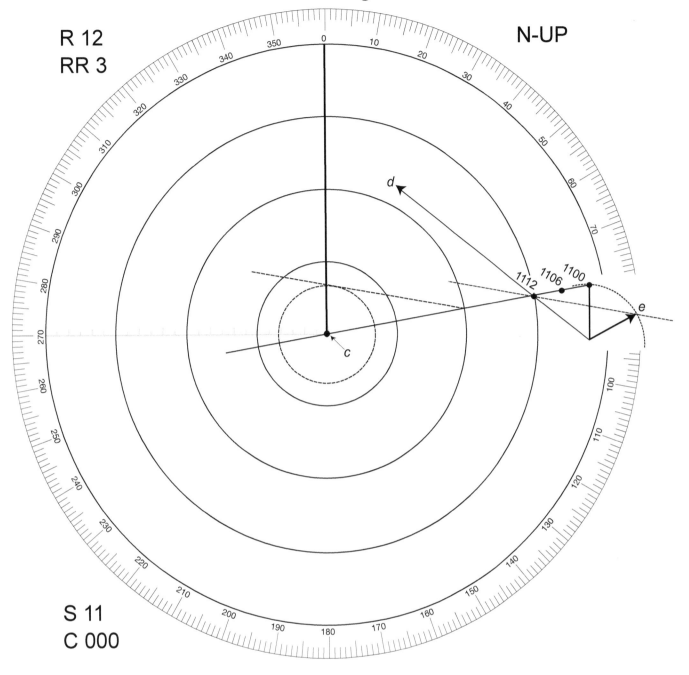

R 12
RR 3

N-UP

S 11
C 000

5. Own ship, on course 220°, speed 12 knots, obtains the following radar bearings and ranges at the times indicated, using a radar range setting of 12 miles:

Time	Bearing	Range (mi.)
0300	297°	11.7
0306	296°	10.0
0312	295°	8.5

Required—

(a) Range at CPA.

(b) Speed of relative movement (SRM); relative speed.

(c) Time at CPA.

(d) True course of contact.

Decision.—When the range to the contact decreases to 6 miles, own ship will change course so that the contact will clear ahead, in minimum time, with a CPA of 3.0 miles.

Required—

(e) New course for own ship.

(f) New SRM after course change.

Answers. Assuming that the contact maintains course and speed: (a) 1.1 nmi, (b) 16.1 kt, (c) 0343, (d) C 162, (e) C 290, (f) 27.8 kt

Plot target at appropriate ranges and bearings; note times. Draw DRM line. Draw CPA line perpendicular to the extended DRM and read range at CPA (a). Measure distance between first and last target contacts and calculate SRM (b). Measure distance to CPA; calculate TCPA (c). Draw 12-minute buoy trail astern from first target plot. Complete the RMD and read true course of contact (d). Draw new DRM from 6-nmi range to pass ahead of own ship and tangent to desired CPA. Parallel the new DRM to intersect most recent target plot. Draw an arc of the buoy tail to intersect the paralleled DRM line. Read new course for own ship (e). Note that the target's actual course and speed line, together with the new DRM line and the arc of the buoy trail complete a new RMD. Measure distance between new buoy trail and 0312 contact, and calculate new SRM (f)

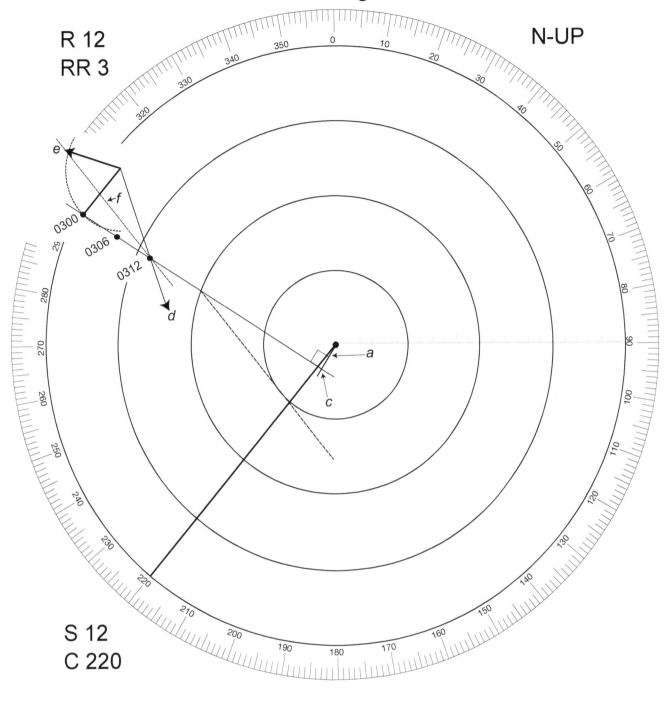

R 12
RR 3

N-UP

S 12
C 220

6. Own ship, on course 316°, speed 21 knots, obtains the following radar bearings and ranges at the times indicated, using a radar range setting of 12 miles:

Time	Bearing	Range (mi.)
1206	357°	11.8
1212	358°	10.2
1218	359°	8.7

Required—

(a) Range at CPA.

(b) Speed of relative movement (SRM); relative speed.

(c) True course of contact.

(d) True speed of contact.

Decision.—When the range to the contact decreases to 6 miles, own ship will change course so that the contact will clear ahead, in minimum time, with a CPA of 3 miles.

Required—

(e) New course for own ship.

Answers. Assuming that the contact maintains course and speed: (a) 1.1 nmi, (b) 15.6 kt, (c) C 269, (d) 12.3 kt, (e) C 003

Plot target at appropriate ranges and bearings; note times. Draw DRM line. Draw CPA line perpendicular to the extended DRM and read range at CPA (a). Measure distance between first and last target contacts; calculate SRM (b). Draw 12-minute buoy trail astern of first target plot. Complete the RMD and read true course of contact (c). Measure distance between buoy trail and 1218 contact; calculate true speed of target (d). Draw new DRM from 6 nmi range ring to pass tangent to desired CPA. Parallel the new DRM to intersect most recent target plot. Draw an arc of the buoy trail to intersect the paralleled new DRM. Read new course for own ship (e).

Radar Maneuvering Problem 6.

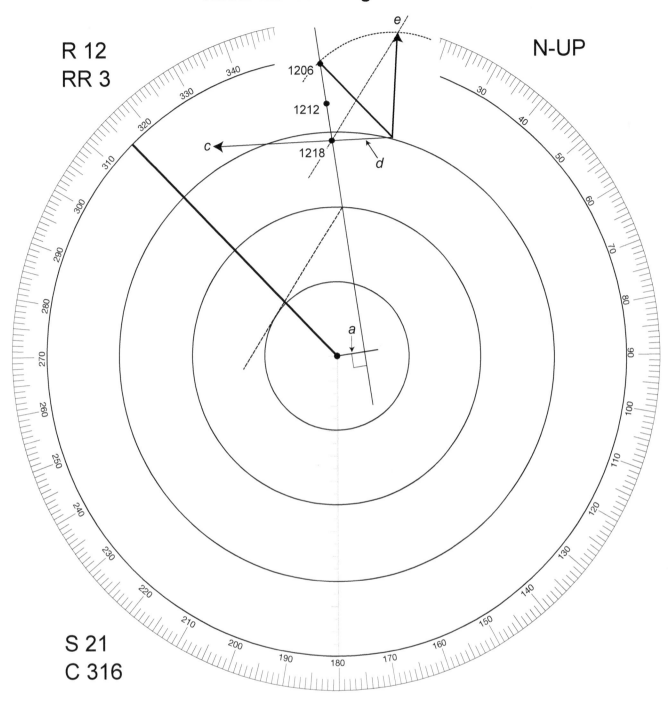

R 12
RR 3

N-UP

1206
1212
1218

c

d

e

a

S 21
C 316

59

7. Own ship, on course 000°, speed 10 knots, obtains the following radar bearings and ranges at the times indicated, using a radar range setting of 12 miles:

Time	Bearing	Range (mi.)
0400	010°	11.1
0406	010°	9.0
0412	010°	7.1

Required—

(a) Range at CPA.

(b) Speed of relative movement (SRM); relative speed.

(c) Time at CPA.

(d) True course of contact.

(e) True speed of contact.

Decision—Own ship will change course at 0418 so that the contact will clear ahead (on own ship's port side), with a CPA of 2 miles.

Required—

(f) New course for own ship.

Answers. Assuming that the contact maintains course and speed: (a) Nil; collision risk exists, (b) 20 kt, (c) 0433, (d) C 200, (e) 10 kt, (f) C 047

Plot target at appropriate ranges and bearings; note times. Draw DRM line; observe that bearing to target is not changing; therefore, range at CPA (a) will be Nil; collision risk exists. Measure distance between 0400 and 0412 contacts; calculate SRM (b). Measure distance to CPA; calculate TCPA (c). Draw 12-minute buoy trail astern from first target plot. Complete the RMD; read true course of contact (d). Measure true speed of target (e). Calculate and draw Turn mark as 6-minute projected location of contact. Draw new DRM from Turn mark to pass tangent to desired CPA range. Parallel the new DRM to intersect most recent target plot. Draw an arc of the buoy trail to intersect the paralleled new DRM. Read new course for own ship (f).

Radar Maneuvering Problem 7.

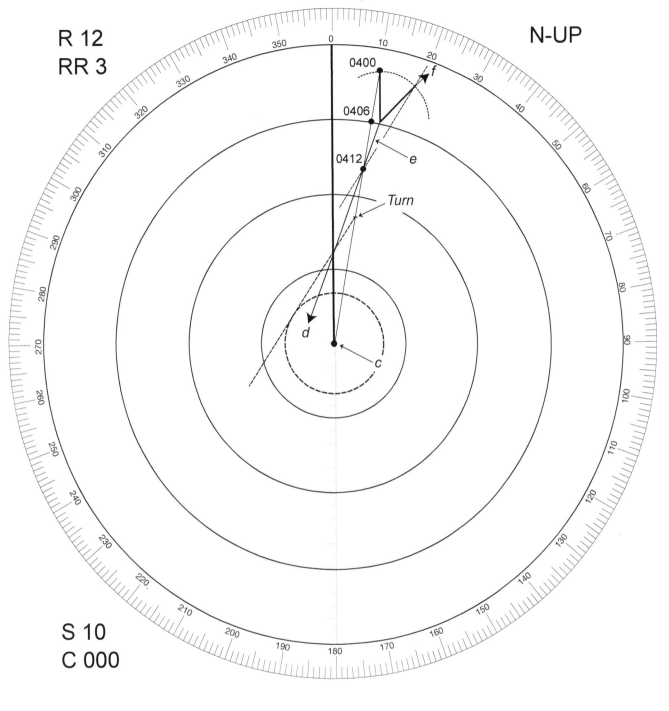

8. Own ship, on course 052°, speed 15 knots, obtains the following radar bearings and ranges at the times indicated, using a radar range setting of 24 miles:

Time	Bearing	Range (mi.)
0340	052°	14.9
0346	052°	11.6
0352	052°	8.3

Required—

(a) Range at CPA.

(b) True course of contact.

(c) Assuming that there are no other vessels in the area and that the contact is a large passenger ship, clearly visible at 0352, is this a crossing, meeting, or overtaking situation?

(d) True speed of contact.

A decision is made to change course when the range to the contact decreases to 6 miles.

Required—

(e) New course of own ship to clear the contact port to port with a CPA of 3 miles.

Answers. Assuming that the contact maintains course and speed: (a) Nil; risk of collision exists, (b) C 232, (c) Meeting, (d) 18 kt, (e) C 119

Plot target at appropriate ranges and bearings; note times. Draw DRM line; observe that bearing of target is not changing; therefore range at CPA (a) will be Nil and collision risk exists. Draw 12-minute buoy trail astern from first target plot. Observe that closure SRM exceeds own ship speed; therefore target must be opposite direction traffic, i.e. a Meeting situation (c). Observe that target true course (b) must be reciprocal of own ship course. Calculate target true speed as the relative distance target covers between 0340 and 0352, minus buoy trail (d). Draw new DRM from 6 nmi range to pass tangent to 3 nmi range ring. Parallel the new DRM to intersect most recent target plot. Draw an arc of the buoy trail to intersect the new DRM. Read new course for own ship (e).

Radar Maneuvering Problem 8.

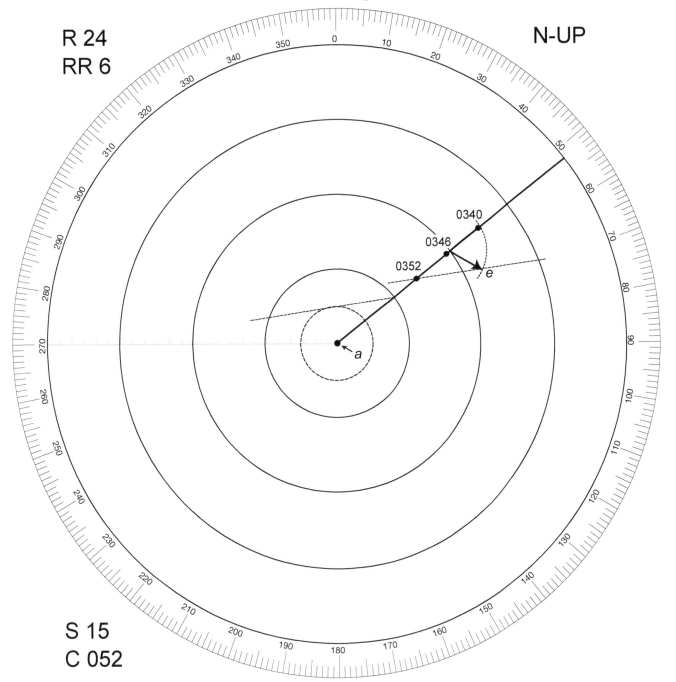

R 24
RR 6

N-UP

S 15
C 052

9. Own ship, on course 070°, speed 16 knots, obtains the following radar bearings and ranges at the times indicated, using a radar range setting of 12 miles:

Time	Bearing	Range (mi.)
0306	015°	10.8
0312	016°	8.3
0318	017°	5.9

Required—

(a) Range at CPA.

(b) Time at CPA.

(c) True course of the contact.

(d) True speed of the contact.

Decision—When the range to the contact decreases to 5 miles, own ship will change speed only so that contact will clear ahead at a distance of 3 miles.

Required—

(e) New speed of own ship.

Answers. Assuming that the contact maintains course and speed: (a) 0.5 nmi, (b) 0332, (c) C 152, (d) 21 kt, (e) 3.1 kt

Plot target at appropriate ranges and bearings; note times. Draw DRM line. Draw CPA line perpendicular to the extended DRM and read range at CPA (a). Measure distance between first and last target contacts; calculate SRM. Measure distance to CPA and calculate TCPA (b). Draw 12-minute buoy trail astern from first target plot. Complete the RMD; read true course of contact (c). Calculate true speed of target (d). Draw new DRM from 5 nmi range to pass tangent to desired CPA. Parallel the new DRM to intersect most recent target plot and thus complete a new RMD. Read the new 12-minute distance on the buoy trail of the RMD; calculate own ship reduced speed (e).

Radar Maneuvering Problem 9.

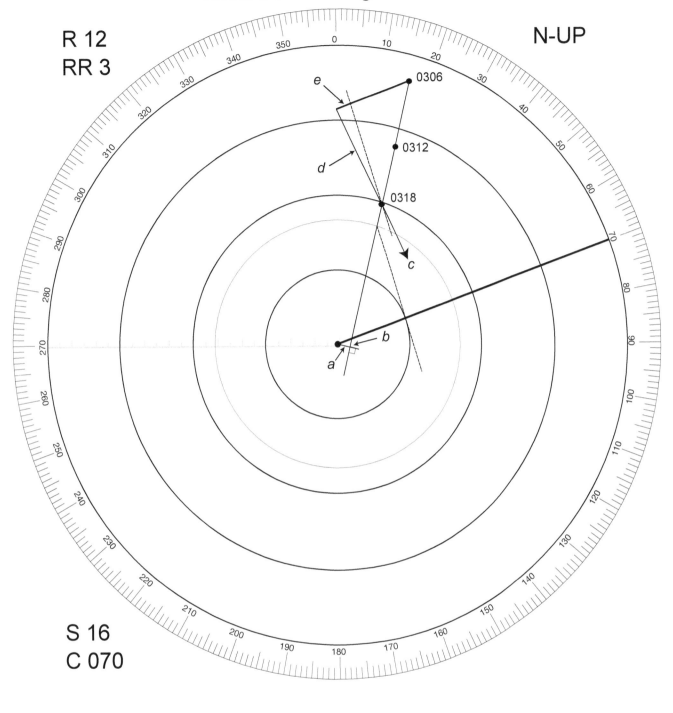

R 12
RR 3

N-UP

S 16
C 070

10. Own ship, on course 093°, speed 18 knots, obtains the following radar bearings and ranges at the times indicated, using a radar range setting of 12 miles:

Time	Bearing	Range (mi.)
0452	112°	5.9
0458	120°	4.2
0504	137°	2.7

Required—

(a) Range at CPA.

(b) Relative distance (MRM) from 0452 to 0504 position of contact.

(c) Speed of relative movement (SRM); relative speed.

(d) Direction of relative movement (DRM).

(e) Distance own ship travels from the time of the first plot (0452) to the time of the last plot (0504) of the contact.

(f) True course and speed of the contact.

Answers. Assuming that the contact maintains course and speed: (a) 1.9 nmi, (b) 3.6 nmi, (c) 18 kt, (d) 273, (e) 3.6 nmi, (f) The contact is either a stationary object or a vessel underway but with no way on.

Plot target at appropriate ranges and bearings; note times. Draw DRM line. Observe that DRM (d) is parallel to own ship course, and is in fact the reciprocal of own ship course. Draw CPA line perpendicular to the extended DRM and read range at CPA (a). Measure distance between first and last contacts (b), and calculate SRM (c). Calculate distance own ship travels in 12 minutes (e). Observe that target SRM = own ship speed; therefore target must be a stationary object or a vessel underway but with no way on (f).

Radar Maneuvering Problem 10.

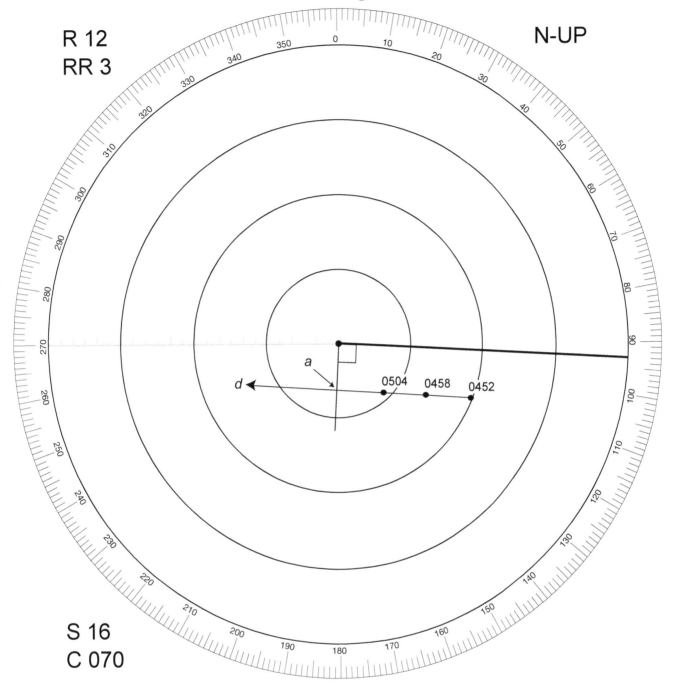

R 12
RR 3

N-UP

S 16
C 070

11. Own ship, on course 315°, speed 11 knots, obtains the following radar bearings and ranges at the times indicated, using a radar range setting of 24 miles:

Time	Bearing	Range (mi.)
0405	319°	17.8
0417	320°	15.6
0429	321°	13.4

Required—

(a) Range at CPA.

(b) True course and speed of the contact.

Decision—When the range to the contact decreases to 8 miles, own ship will change course so that the contact will pass safely to starboard with a CPA of 3 miles.

Required—

(c) New course for own ship.

Answers. Assuming that the contact maintains course and speed: (a) 1.9 nmi, (b) The contact is either stationary or a vessel with little or no way on, or own ship is being set by current. (c) C 307

Problem 11. Plot target at appropriate ranges and bearings; note times. Draw DRM line. Draw line perpendicular to the extended DRM and read range at CPA (a). Draw 24-minute buoy trail astern from first target plot. Complete RMD and calculate target true course and speed. Observe that buoy trail almost overlaps DRM; therefore target must be stationary or a vessel with little or no way on (b). Draw new DRM from 8 nmi range point to pass tangent to 3 nmi range ring. Parallel the new DRM to intersect 0429 contact. Draw an arc of the buoy trail to intersect new DRM and read new course for own ship (c).

Radar Maneuvering Problem 11.

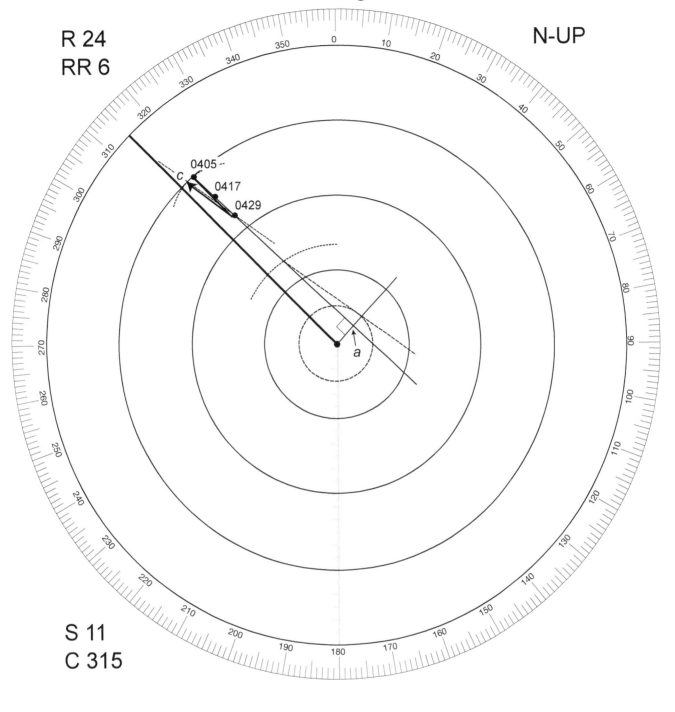

R 24
RR 6

N-UP

S 11
C 315

12. Own ship, on course 342° speed 11 knots, (half speed), obtains the following radar bearings and ranges at the times indicated, using a radar range setting of 12 miles:

Time	Bearing	Range (mi.)
0906	287°	12.0
0912	287°	10.2
0918	288°	8.4

Required—

(a) Range at CPA.

(b) True course of the contact.

(c) True speed of the contact.

(d) Is this a crossing, meeting, or overtaking situation?

Own ship is accelerating to full speed of 18 knots.

Decision—Own ship will change course at 0924 when the speed is 15 knots so that the contact will clear astern with a CPA of 2 miles.

Required—

(e) New course for own ship.

Answers. Assuming that the contact maintains course and speed: (a) 0.8 nmi, (b) 065, (c) 15.7 kt, (d) Crossing, (e) C 011

Plot target at appropriate ranges and bearings; note times. [Assume 0912 and 0918 to be the more accurate contacts.] Draw DRM line. Draw CPA line perpendicular to the extended DRM and read range at CPA (a). Draw 12 minute buoy trail astern from first target plot. Complete the RMD and read true course of contact (b) and calculate true speed of contact (c). Observe that this is a Crossing situation (d). Establish 0924 Turn point by calculation [assume 15 kts]. Draw new DRM to pass tangent to desired CPA. Parallel the new DRM to intersect most recent target plot. Draw an arc of the buoy trail using new speed [assume 15 kts], and position to complete the new RMD. Read new course for own ship (e).

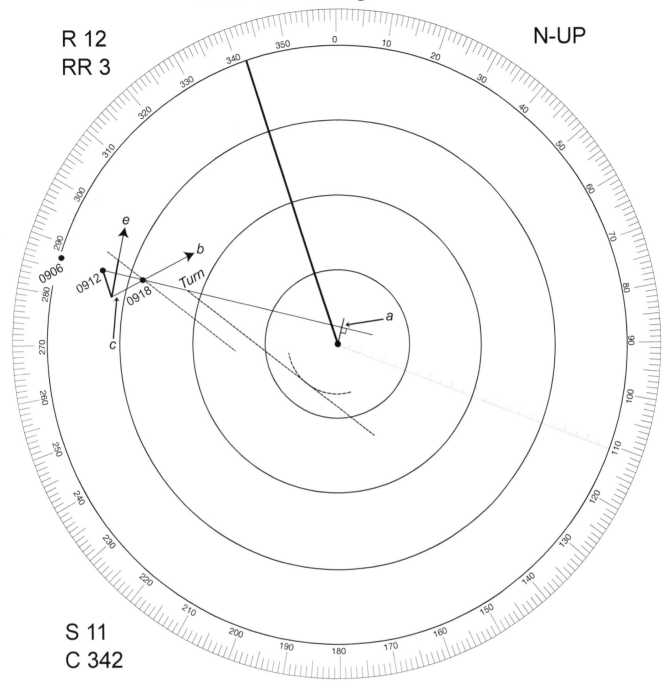

R 12
RR 3

N-UP

0906

0912

0918

Turn

e

b

c

a

S 11
C 342

13. Own ship, on course 350°, speed 18 knots, obtains the following radar bearings and ranges at the times indicated, using a radar range setting of 12 miles:

Time	Bearing	Range (mi.)
0200	030°	10.0
0203	029°	8.7
0206	028°	7.4

Required—

(a) Range at CPA.

(b) True course of the contact.

(c) True speed of the contact.

When the range to the contact decreases to 6 miles, own ship changes course to 039°.

(d) New range at CPA.

(e) Describe how the new time at CPA would be computed.

(f) New time at CPA.

(g) At what bearing and range to the contact can own ship safely resume the original course of 350° and obtain a CPA of 3 miles?

(h) What would be the benefit, if any, of bringing own ship slowly back to the original course of 350° once the point referred to in (g) above is reached?

Answers. Assuming that the contact maintains course and speed: (a) 1.0 nmi, (b) C 259, (c) 18.7 kt, (d) 3.3 nmi, (e) Determine the original SRM; calculate time at Turn point. Draw new buoy trail, copy target trail, complete the RMD. Determine new SRM; then using it, determine ETE to new CPA. Convert to TCPA. (f) 0218, (g) When the contact bears 330, range 3.3 miles. (h) The slow return to the original course will serve to insure that the contact will remain outside the 3-mile danger or buffer zone after own ship is steady on C 350.

Plot target at appropriate ranges and bearings; note times. Draw DRM line. Draw line perpendicular to the extended DRM and read range at CPA (a). Calculate SRM and use to determine time at 6 nmi Turn point. Draw 6 minute buoy trail astern from first target plot. Complete the RMD and read true course of contact (b). Measure and calculate true speed of target (c). Draw an arc of the buoy trail to new course, keeping length the same. Complete the new RMD, yielding the new DRM. Measure and calculate new SRM. Parallel the new DRM to intersect the Turn point. Determine new range at CPA (d). Determine (e) new TCPA using new SRM and distance remaining from Turn point. Calculate new TCPA (f). Return to original course after target passes CPA (g). The slow return to the original course will insure (h) that contact remains outside the 3-mile safety zone after own ship is steady on 350.

Radar Maneuvering Problem 13.

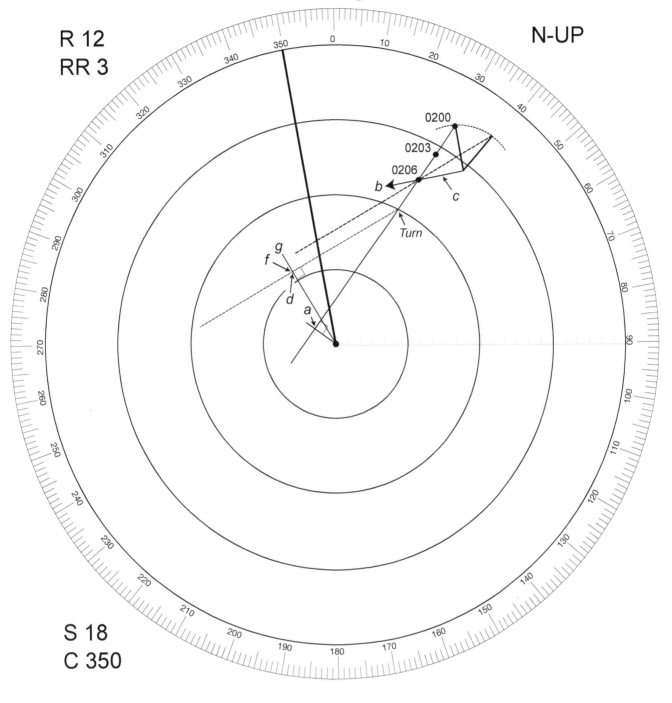

S 18
C 350

73

14. Own ship, on course 330°, speed 20 knots, obtains the following bearings and ranges at the times indicated, using a radar range setting of 12 miles:

Time	Bearing	Range (mi.)
0608	300°	12.0
0614	300°	10.0
0620	300°	8.0

Required—

(a) Range at CPA.

(b) Time at CPA.

(c) True course of the contact.

(d) True speed of the contact.

(e) What danger, if any, would be present if own ship maintained course and speed and contact changed course to 120° at 0620?

Assume that the contact maintains its original course and speed and that own ship's speed has been reduced to 11.5 knots when the range to the contact has decreased to 6 miles.

Required—

(f) New range at CPA.

(g) Will the contact pass ahead or astern of own ship?

Answers. (a) Nil; risk of collision exists, (b) 0644, (c) C 045, (d) 10.3 kt. (e) None; contact would pass abeam at 2.7 nmi CPA, (f) 1.9 nmi, (g) Ahead, at 2.5 nmi

Plot target at appropriate ranges and bearings; note times. Observe DRM line indicates collision risk (a). Calculate SRM; then calculate TCPA (b). Draw 12 min buoy trail astern from first target plot. Complete the RMD and read true course of contact (c). Measure and calculate the true speed of contact (d). Draw an arc of the target vector and complete the new RMD to obtain a new DRM line. Parallel DRM line to 0620 contact and extend to CPA. Observe that target will pass abeam at a safe distance (e). From the 6 nmi Turn point, draw the new 11.5 kt buoy trail and contact's true speed and direction vector [the two known sides of a new RMD]. Complete the RMD with the new DRM line. Observe target will pass ahead at a safe distance (g). Determine new CPA (f).

Radar Maneuvering Problem 14.

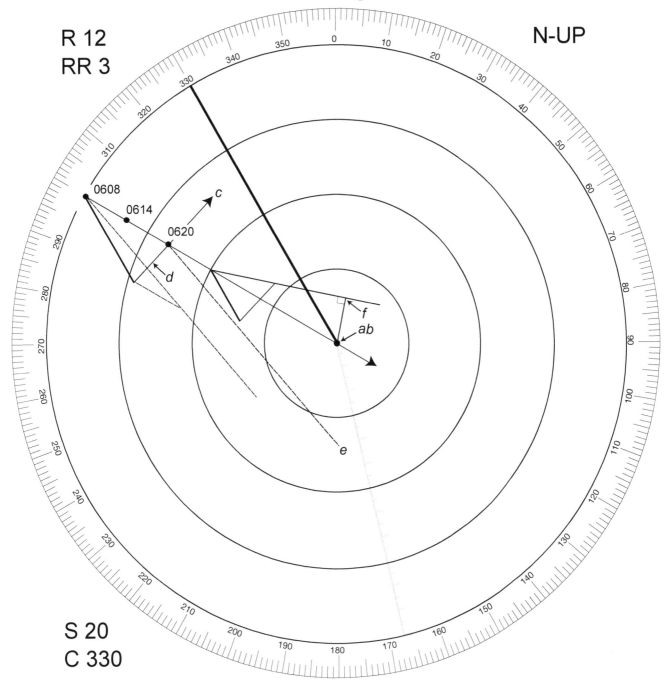

S 20
C 330

0608
0614
0620

c
d
e
f
ab

15. Own ship, on course 022°, speed 32 knots, obtains the following radar bearings and ranges at the times indicated, using a radar range setting of 24 miles:

Time	Contact A	Contact B	Contact C
0423	070°-23.2 mi.	170°-23.8 mi.	025°-22.6 mi.
0426	070°-21.1 mi.	170°-23.8 mi.	023°-21.2 mi.
0429	070°-19.1 mi.	170°-23.8 mi.	020°-19.0 mi.

The observations are made on a warm, summer morning. The weather is calm; the sea state is 0. From sea water temperature measurements and weather reports, it is determined that the temperature of the air immediately above the sea is 12° F cooler than the air 300 feet above the ship. Also, the relative humidity immediately above the sea is 30% greater than at 300 feet above the ship.

Required—

(a) Since the contacts are detected at ranges longer than normal, to what do you attribute the radar's increased detection capability?

(b) Ranges at CPA for the three contacts.

(c) True courses of the contacts.

(d) True speeds of the contacts.

(e) Which contact presents the greatest threat?

(f) If own ship has adequate sea room, should own ship come left or right of contact A?

Decision—When the range to contact A decreases to 12 miles, own ship will change course so that no contact will pass within 4 miles.

Required—

(g) New course for own ship.

Answers. Assuming that the contacts maintain course and speed: (a) Super-refraction, (b) Contact A nil; Contact, 23.8 nmi; Contact C 9.3 nmi, (c) Contact A 301; Contact B 022; Contact C 280, (d) Contact A 31 kt; Contact B 32 kt; Contact C 19 kt, (e) Contact A; it is on collision course, (f) Come right, (g) C 059

Plot targets at appropriate ranges and bearings; note times. Draw DRM lines for all. Measure ranges at CPA (b). Observe that Contact A's DRM indicates collision risk. Solve RMD for each applicable target to determine actual course (c) and speed (d) of Contacts A & C. Observe that Contact B is maintaining same course and speed as own ship. Observe that optimum avoidance strategy is to come right to pass astern (f) of Contact A. Draw desired DRM from Contact A's DRM at the 12-mile range ring to clear by the desired CPA. Parallel the DRM line to intersect the most recent contact; then, using the now known true course and speed vector of the target, draw an arc of the buoy trail to complete a new RMD. Read the new course for own ship (g). Calculate Contact A's ETE to Turn. Use this time to calculate Turn point for Contact C. Complete RMDs for the other targets based on the new buoy trail to confirm no new conflict has been created.

Radar Maneuvering Problem 15.

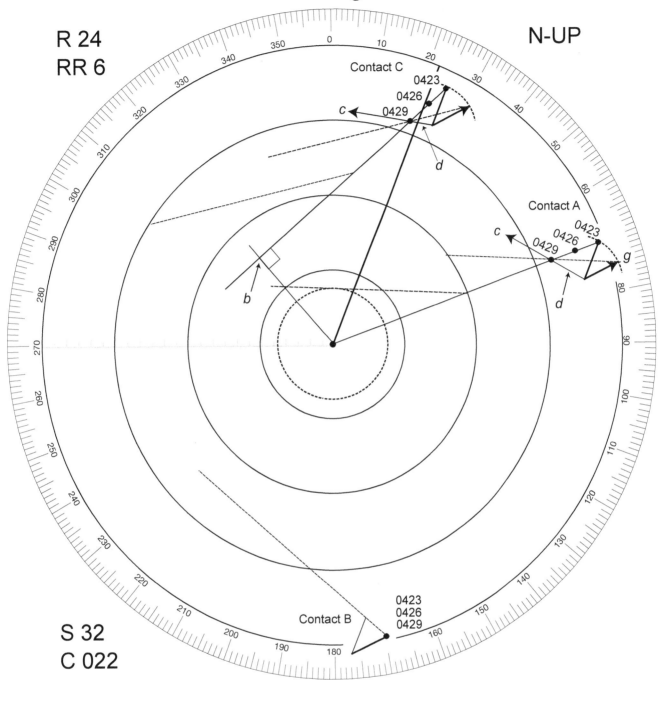

R 24
RR 6

N-UP

S 32
C 022

77

16. Own ship, on course 120°, speed 12 knots, obtains the following radar bearings and ranges at the times indicated, using a radar range setting of 12 miles:

Time	Contact A	Contact B	Contact C
0300	095°-8.7 mi.	128°-10.0 mi.	160°-7.7 mi.
0306	093°-7.8 mi.	128°-8.3 mi.	164°-7.0 mi.
0312	090°-7.0 mi.	128°-6.6 mi.	170°-6.3 mi.

Required—

(a) Ranges at CPA for the three contacts.

(b) True courses of the contacts.

(c) Which contact presents the greatest danger?

(d) Which contact, if any, might be a lightship at anchor?

Decision—

When the range to contact B decreases to 6 miles, own ship will change course to 190°.

Required—

(e) At what time will the range to contact B be 6 miles?

(f) New CPA of contact C after course change to 190°.

Answers. Assuming the contacts maintain course and speed: (a) Contact A 2.9 nmi; contact B - nil; contact C 4.5 nmi, (b) Contact A 137; contact B 326; contact C 107, (c) Contact B; it is on collision course, (d) None, (e) 0314, (f) 3.2 nmi

Plot targets at appropriate ranges and bearings; note times. Draw DRM lines for all. Measure ranges at CPA (a). Observe that Contact B's DRM indicates collision risk (c). Using 12-minute buoy trail, solve RMD for each target and read true courses (b). Observe that no contact has a DRM line that matches buoy trail; therefore all are underway; none can be a lightship under anchor (d). Compute Contact B ETA at 6 nmi range Turn point. Determine Contact C's projected Turn point. Draw two sides of a new RMD [the now-known true course and speed of Contact C, plus own ship's new buoy trail]. Complete the RMD with a new DRM line. Measure CPA (f).

Radar Maneuvering Problem 16.

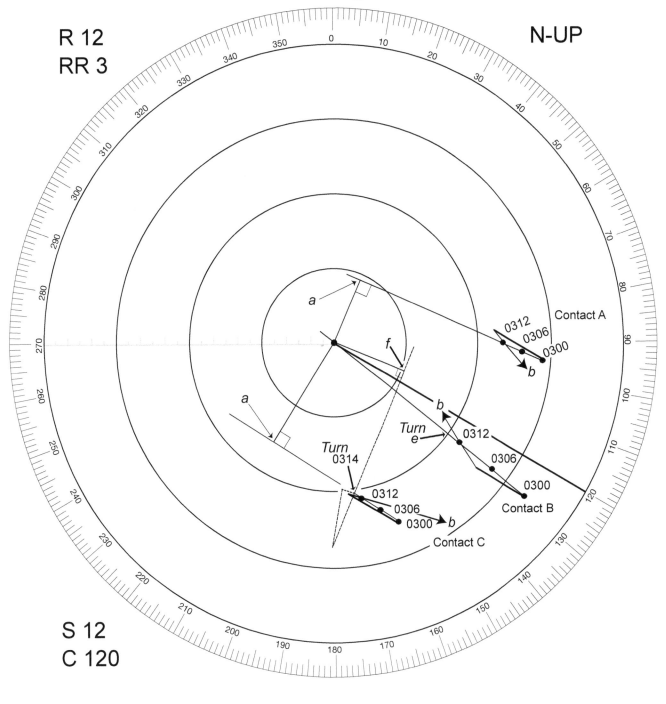

S 12
C 120

79

17. Own ship, on course 298°, speed 13 knots, obtains the following radar bearings and ranges at the times indicated, using a radar range setting of 20 miles:

Time	Bearing	Range (mi.)
0639	267°	19.0
0651	266.5°	16.0
0709	265°	11.5
0729	261°	6.5
0735	255.5°	4.9
0737	252°	4.3
0741	242.5°	3.3

Required—

(a) Range at CPA as determined at 0729.

(b) Time at CPA as determined at 0729.

(c) Course of other ship as determined at 0729.

(d) Speed of other ship as determined at 0729.

(e) Range at CPA as determined at 0741.

(f) Time at CPA as determined at 0741.

(g) Course of other ship as determined at 0741.

(h) Speed of other ship as determined at 0741.

Answers. (a) 1.0 nmi, (b) 0754, (c) C 031, (d) 7.0 kt, (e) 1.9 nmi, (f) 0749, (g) C 065, (h) 7.5 kt

Plot target at appropriate ranges and bearings; note times. Draw DRM lines for target, including that resulting from course change at 0729. Measure ranges at CPA (a) and (e). Calculate SRM for target on both DRM; then calculate both TCPA (b) and (f). Draw 20-minute buoy trail commencing at 0709 contact. Complete RMD through 0729 contact. Read true course (c) and calculate speed (d) of contact. Draw 6-minute buoy trail commencing at 0735 contact. Complete RMD through 0741 contact. Measure true course (g) and speed (h) of contact.

80

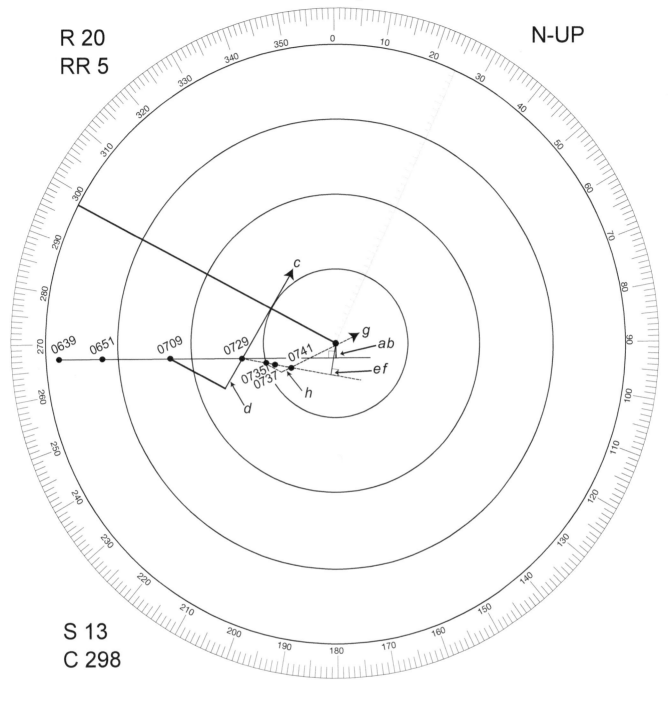

R 20
RR 5

N-UP

S 13
C 298

81

18. Own ship, on course 073°, speed 19.5 knots, obtains the following radar bearings and ranges at the times indicated, using a radar range setting of 20 miles:

Time	Bearing	Range (mi.)
1530	343°	16.2
1540	343°	14.7
1546	343°	13.8
1558	343°	12.0
1606	342.5°	10.9
1612	341.5°	10.1
1624	339.5°	8.4
1632.5	336°	7.3
1644	328.5°	6.0
1657	315°	4.7

Required—

(a) Range at CPA as determined at 1558.

(b) Time at CPA as determined at 1558.

(c) Course of other ship as determined at 1558.

(d) Speed of other ship as determined at 1558.

(e) Range at CPA as determined at 1624.

(f) Time at CPA as determined at 1624.

(g) Course of other ship as determined at 1624.

(h) Speed of other ship as determined at 1624.

(i) Range at CPA as determined at 1657.

(j) Time at CPA as determined at 1657.

(k) Course of other ship as determined at 1657.

(l) Speed of other ship as determined at 1657.

Answers. (a) nil; collision risk exists, (b) 1718, (c) C 098, (d) 21.5 kt, (e) 1.5 nmi, (f) 1721, (g) C 098, (h) 20.3 kt, (i) 3.7 nmi, (j) 1719, (k) C 098, (l) 18 kt

Plot target at appropriate ranges and bearings; note times. Draw DRM lines for target at the requested times, indicating target maneuvering. Measure ranges at CPA (a), (e) and (i). Calculate SRM at the requested times; then calculate TCPA (b), (f) and (j). Draw 12-minute buoy trails commencing at contact times 1558 and 1612. Complete RMDs to determine target true course (c) and (g), and speed (d) and (h). Draw 13 minute buoy trail commencing at contact time 1644. Complete RMD to determine target true course (k) and speed (l).

Radar Maneuvering Problem 18.

R 20
RR 5

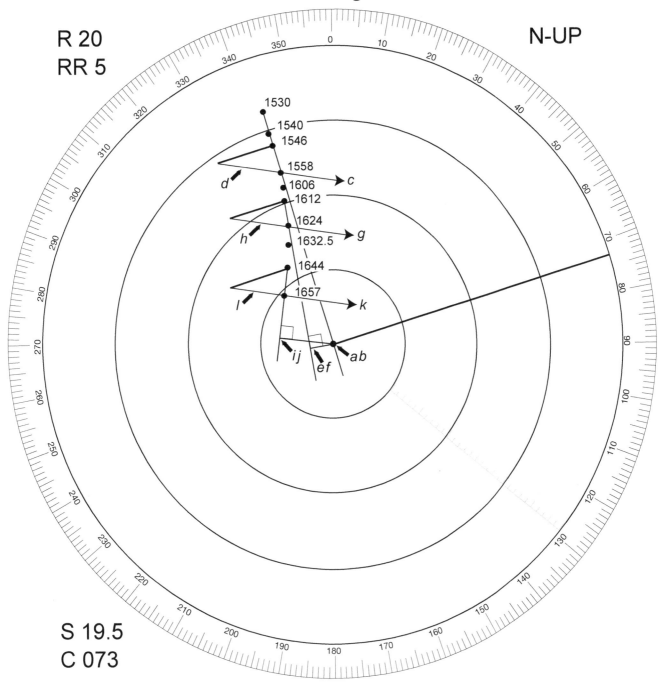

1530
1540
1546
1558
1606
1612
1624
1632.5
1644
1657

c
d
g
h
k
l
ij
ef
ab

S 19.5
C 073

19. Own ship, on course 140°, speed 5 knots, obtains the following radar bearings and ranges at the times indicated, using a radar range setting of 12 miles:

Time	Bearing	Range (mi.)
0257	142°	10.5
0303	141.5°	8
0308	141°	6
0312	135°	4.5
0314	126.5°	4
0317	110.5°	3.2

Required—

(a) Range at CPA as determined at 0308.

(b) Time at CPA as determined at 0308.

(c) Course of other ship as determined at 0308.

(d) Speed of other ship as determined at 0308.

(e) Range at CPA as determined at 0317.

(f) Time at CPA as determined at 0317.

(g) Course of other ship as determined at 0317.

(h) Speed of other ship as determined at 0317.

Answers. (a) 0.25 nmi, (b) 0323, (c) C 324, (d) 19.6 kt, (e) 2.9 nmi, (f) 0321, (g) C 002, (h) 20.9 kt

Plot target at appropriate ranges and bearings; note times. Draw DRM line and measure range at CPA (a). Calculate SRM; then calculate TCPA (b). Draw 5-minute buoy trail commencing at contact 0303. Complete RMD to determine target true course (c) and speed (d). Draw DRM line commencing at contact 0312. Note that position of 0314 contact appears to be an anomaly. Measure range at CPA (e). Calculate SRM; then calculate TCPA (f). Draw 5-minute buoy trail commencing at 0312 contact. Complete RMD to determine target's actual course (g) and speed (h).

Radar Maneuvering Problem 19.

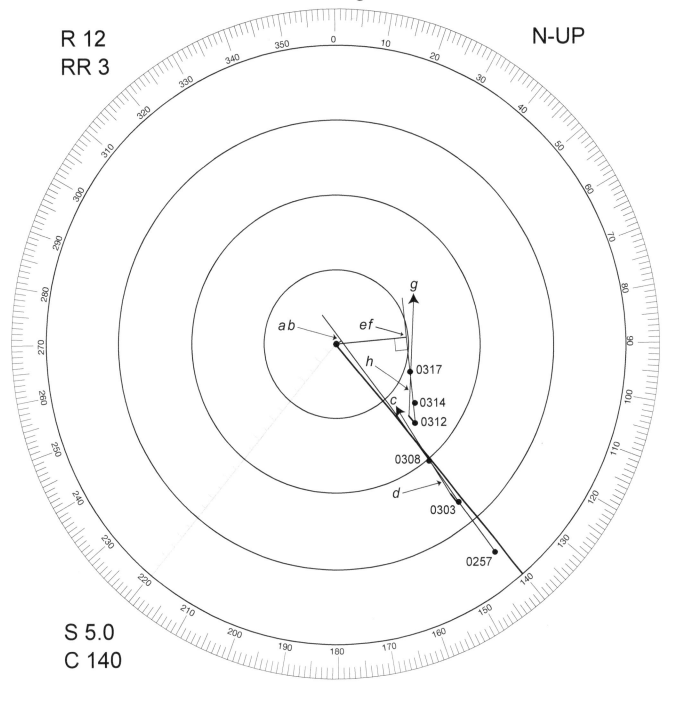

R 12
RR 3

N-UP

S 5.0
C 140

85

20. Own ship, on course 001°, speed 15 knots, obtains the following radar bearings and ranges at the times indicated, using a radar range setting of 15 miles:

Time	Bearing	Range (mi.)
2243	138°	14.0
2255	137.5°	12.6
2318	136°	9.9
2332	140°	8.0
2351	166.5°	5.5
0002.5	191.5°	5.0
0008	204°	5.1
0014	214°	5.1
0020	222°	4.95
0026	230°	4.85

Required—

(a) Range at CPA as determined at 2318.

(b) Time at CPA as determined at 2318.

(c) Course of other ship as determined at 2318.

(d) Speed of other ship as determined at 2318.

(e) Predicted range of other vessel as it crosses dead ahead of own ship as determined at 2318.

(f) Predicted time of crossing ahead as determined at 2318.

(g) Course of other ship as determined at 2351.

(h) Speed of other ship as determined at 2351.

(i) Predicted range of other vessel as it crosses dead astern of own ship as determined at 2351.

(j) Predicted time of crossing astern as determined at 2351.

(k) Direction of relative movement between 0002.5 and 0008.

(l) Relative speed between 0002.5 and 0008.

(m) Course of other ship as determined at 0026.

(n) Speed of other ship as determined at 0026.

Answers. (a) 1.2 nmi, (b) 0042, (c) C 349, (d) 21.0 kt, (e) 1.9 nmi, (f) 0055, (g) C 326, (h) 21.0 kt, (i) 5.1 nmi, (j) 2358, (k) 282.6, (l) 12.0 kt, (m) C 350, (n) 21.0 kt

Plot target at appropriate ranges and bearings; note times. Draw DRM lines for target at the requested times, indicating target maneuvering. Measure ranges at CPA (a) and (b). Draw 23-minute buoy trail commencing at 2255 contact; then complete RMD to determine target actual course (c) and speed (d). Measure predicted range at crossing ahead of own ship (e). Calculate SRM; then calculate time of target crossing ahead of own ship (f). Draw 19-minute buoy trail commencing at 2332 contact. Complete RMD to determine actual course (g) and speed (h) of contact. Extend heading line down screen; Calculate SRM and time of target crossing astern (i). Measure range of target as she crosses astern (j). Draw SRM between 0002.5 and 0008 contacts to determine DRM (k) and SRM (l). Draw 6-minute buoy trail commencing 0020 contact; then complete RMD to determine actual course (m) and speed (n) of target.

Radar Maneuvering Problem 20.

N-UP

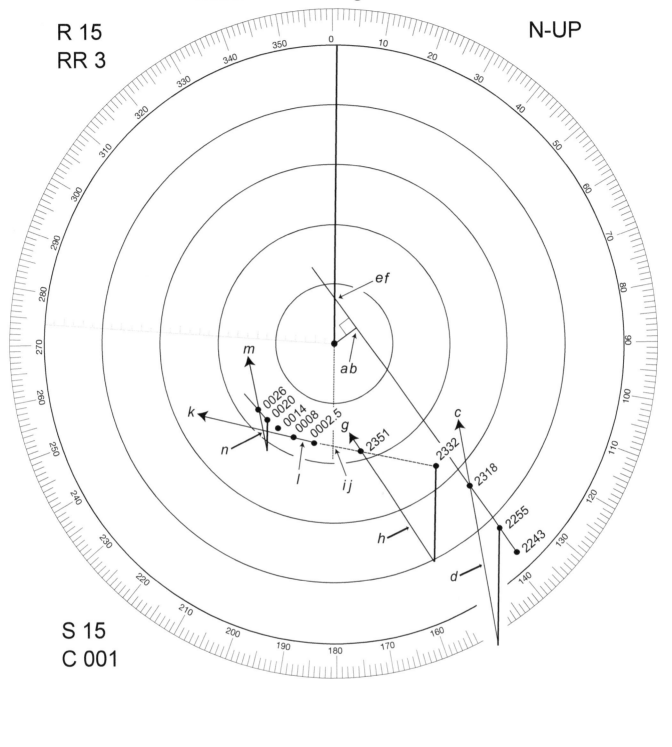

21. Own ship, on course 196°, speed 8 knots, obtains the following radar bearings and ranges at the times indicated, using a radar range setting of 12 miles:

Time	Bearing	Range (mi.)
2303	016°	11.0
2309	016°	10.0
2318	016°	8.5
2330	016°	6.5
2340	011.5°	4.9
2350	359.5°	3.4
2400	333.5°	2.2
0010.5	286°	2.0
0020	247.5°	2.5
0026	233.5°	3.2

Required—

(a) Range at CPA as determined at 2318.

(b) Time at CPA as determined at 2318.

(c) Course of other ship as determined at 2318.

(d) Speed of other ship as determined at 2318.

(e) Range at CPA as determined at 2400.

(f) Time at CPA as determined at 2400.

(g) Course of other ship as determined at 2400.

(h) Speed of other ship as determined at 2400.

(i) Course of other ship as determined at 0026.

(j) Speed of other ship as determined at 0026.

Answers. (a) nil; collision risk exists, (b) 0009, (c) C 196, (d) 18.0 kt, (e) 1.9 nmi, (f) 0006, (g) C 206, (h) 18.0 kt, (i) C 196, (j) 18.0 kt

Plot target at appropriate ranges and bearings indicating target maneuvering; note times. Draw DRM; note constant bearing and risk of collision (a). Calculate SRM; then calculate TCPA (b). Draw 6-minute buoy trail commencing 2303 contact. Observe buoy trail and DRM are in alignment; therefore, target is an overtaking vessel on same course as own ship (c). Calculate actual speed of target (d) as sum of buoy trail and 2303 – 2309 contact distance, converted to knots. Draw DRM from 2350 contact; then measure range at CPA (e). Calculate SRM; then calculate TCPA (f). Draw 10-minute buoy trail commencing 2350 contact. Complete RMD to determine target actual course (g) and speed (h). Draw DRM from 0020 contact; observe alignment same as own ship, and thus target actual course (i). Draw 6-minute buoy trail commencing 0020 contact; sum distance of buoy trail and 0020 – 0026 contact; then convert to knots to determine target actual speed (j).

Radar Maneuvering Problem 21.

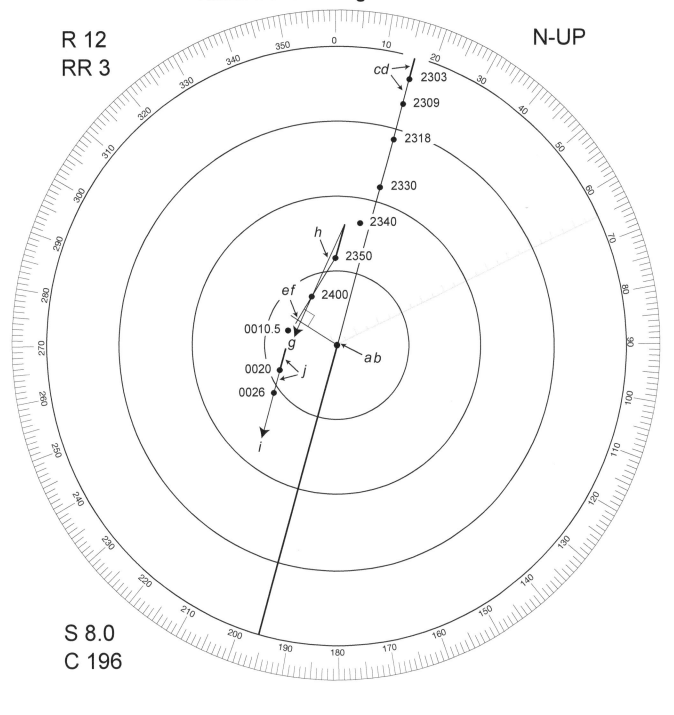

R 12
RR 3

N-UP

S 8.0
C 196

22. Own ship, on course 092°, speed 12 knots, obtains the following radar bearings and ranges at the times indicated, using a radar range setting of 16 miles:

Time	Bearing	Range (mi.)
1720	335°	15.0
1750	334.5°	11.7
1830	333°	7.2
1854	325.5°	4.5
1858	315.5°	4.0
1902	303.5°	3.6
1906	289.5°	3.4
1914	263.5°	3.3
1930	212.5°	3.8
1950	184.5°	6.8

Required—

(a) Range at CPA as determined at 1830.

(b) Time at CPA as determined at 1830.

(c) Course of other ship as determined at 1830.

(d) Speed of other ship as determined at 1830.

(e) Course of other ship as determined at 1906.

(f) Speed of other ship as determined at 1906.

(g) Course of other ship as determined at 1950.

(h) Speed of other ship as determined at 1950.

Answers. (a) 0.5 nmi, (b) 1934, (c) C 114, (d) 16.1 kt, (e) C 146, (f) 15.9 kt, (g) C 124, (h) 19.9 kt

Plot target at appropriate ranges and bearings indicating target maneuvering; note times. Draw DRM. Determine range at CPA (a). Calculate SRM; then calculate TCPA (b). Draw 40-minute buoy trail commencing 1750 contact. Complete RMD to determine target actual course (c) and speed (d). Draw 4-minute buoy trail commencing 1902 contact. Complete RMD to determine target actual course (e) and speed (f). Draw 20-minute buoy trail commencing 1930 contact. Complete RMD to determine target actual course (g) and speed (h).

Radar Maneuvering Problem 22.

R 16
RR 4

S 12
C 092

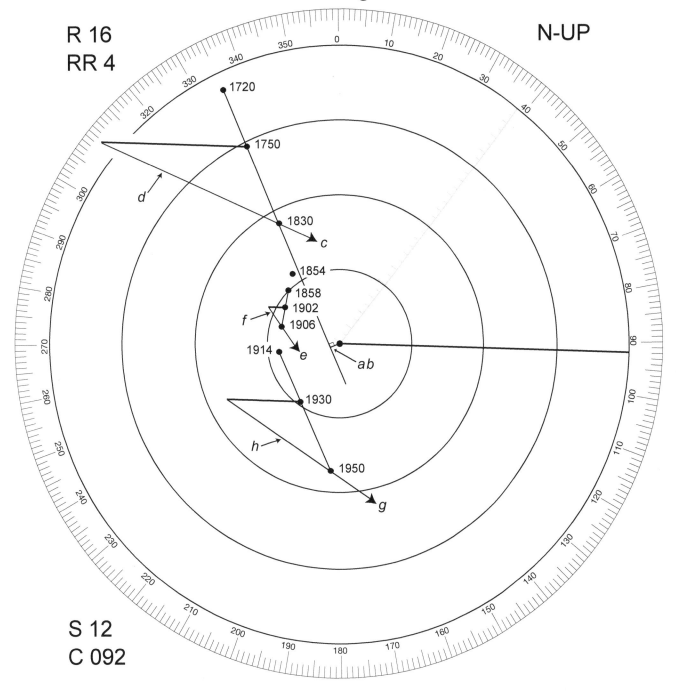

23. Own ship, on course 080°, speed 12.5 knots, obtains the following radar bearings and ranges at the times indicated, using a radar range setting of 16 miles:

Time	Bearing	Range (mi.)
0035	038°	14.5
0044	038.5°	13.2
0106	040°	10.0

Required—

(a) Range at CPA.

(b) Time at CPA.

(c) Course of other ship.

(d) Speed of other ship.

Decision—When the range decreases to 8.0 miles, own ship will turn to the left to increase the CPA distance to 3.0 miles.

Required—

(e) Predicted time of change of course.

(f) Predicted bearing of other ship when own ship changes course.

(g) New course for own ship.

(h) Time at new CPA.

(i) Time at which own ship is dead astern of other ship.

Answers. (a) 1.0 nmi, (b) 0214, (c) C 124, (d) 9.0 kt, (e) 0120, (f) 041.5, (g) C 064, (h) 0200, (i) 0204

Plot target at appropriate ranges and bearings; note times. Draw DRM. Determine range at CPA (a). Measure SRM; then calculate TCPA (b). Draw 22-minute buoy trail commencing 0044 contact. Complete RMD to determine target actual course (c) and speed (d). Use SRM to calculate time at turn point (e). Measure bearing to turn point (f). Draw new DRM to pass 3 nmi CPA. Parallel new DRM to convenient open area. Draw a line representing target's now known actual course and speed (can copy target's 22-minute value from above) adjacent to new DRM. Copy 22-minute buoy trail and position to pivot from astern of target's actual course and speed line. Swing buoy trail until other end is adjacent to DRM line. Read new course for own ship (g). Using new SRM calculate TCPA (h). Draw target course line extending from own ship; calculate time of passing astern of target (i).

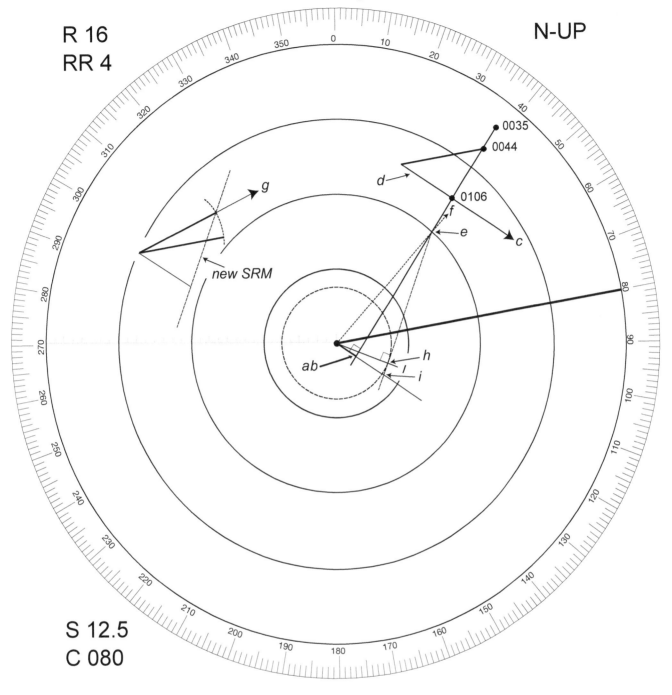

R 16
RR 4

N-UP

new SRM

0035
0044
0106

g
d
f
e
c
ab
h
l
i

S 12.5
C 080

24. Own ship, on course 251°, speed 18.5 knots, obtains the following radar bearings and ranges at the times indicated, using a radar range setting of 20 miles:

Time	Bearing	Range (mi.)
0327	314°	16.2
0337	314.5°	14.7
0351	315°	12.6
0401	315.5°	11.1
0413.5	315°	9.1
0422	305°	6.7

Required—(As determined at 0401.)

(a) Range at CPA.

(b) Time at CPA.

(c) Course of other ship.

(d) Speed of other ship.

Decision—Own ship will pass astern of other vessel, with a CPA of 4.0 miles and new direction of relative movement perpendicular to own ship's original course, maintaining a speed of 18.5 knots. The original course will be resumed when the other ship is dead ahead of this course.

Required—

(e) New direction of relative movement.

(f) Predicted time of change of course.

(g) Predicted bearing of other ship when own ship changes course.

(h) Predicted range of other ship when own ship changes course.

(i) New course for own ship.

(j) Predicted new relative speed.

(k) Predicted time at which other ship is dead ahead of own ship.

(l) Predicted range of other ship when it is dead ahead of own ship.

(m) Predicted time at CPA, as determined at 0422.

(n) Bearing of other ship when it is dead ahead of own ship's original course.

(o) Predicted time of resuming original course.

Answers. (a) 0.75 nmi, (b) 0515, (c) C 222, (d) 16.1 kt, (e) 161, (f) 0412, (g) 316.1, (h) 9.5 nmi, (i) C 291, (j) 19.8 kt, (k) 0428, (l) 5.2 nmi, (m) 0438.5, (n) 251, (o) 0438.5

Plot target at appropriate ranges and bearings; note times. Draw DRM. Determine range at CPA (a). Measure SRM; then calculate TCPA (b). Draw 10-minute buoy trail commencing 0351 contact. Complete RMD to determine target actual course (c) and speed (d). Draw 4 nmi range ring. Draw new DRM perpendicular to heading line, to intercept DRM. Observe new DRM (e). Measure and calculate SRM, measure distance to turn, calculate time at turn (f). Measure bearing to turn point (g). Measure range to turn point (h). Parallel new DRM to convenient open area. Draw a line representing target's now known actual course and speed (can copy target's 10-minute value

Radar Maneuvering Problem 24.

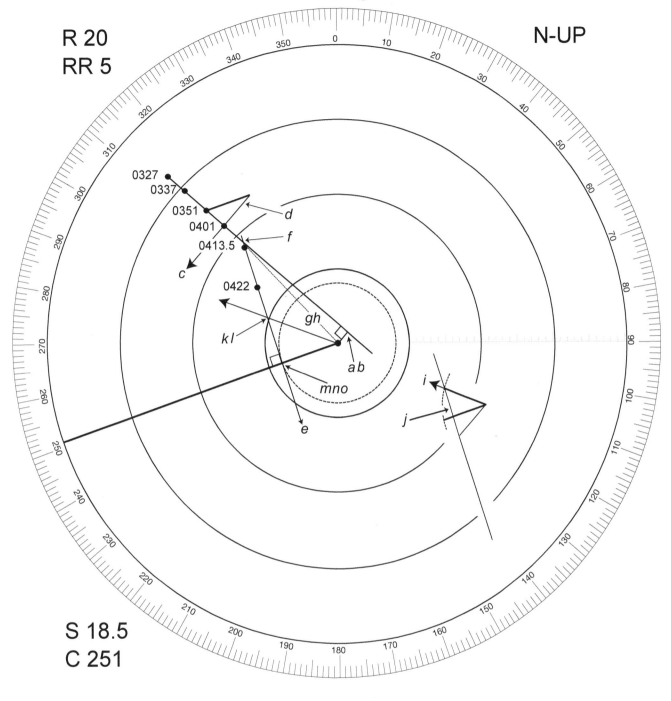

S 18.5
C 251

from above) adjacent to new DRM. Copy 10-minute buoy trail and position to pivot from astern of target's actual course and speed line. Swing buoy trail until other end is adjacent to DRM line. Read new course for own ship (i). Measure and calculate new SRM (j). Draw new course line extending from own ship. Measure and calculate time target crosses ahead of own ship (k). Measure range to target crossing ahead (l). Calculate TCPA (m). Observe bearing of target when ahead of own ship's original course (n). Observe time crossing ahead of original course is same as TCPA (o).

25. Own ship, on course 035°, speed 20 knots, obtains the following radar bearings and ranges at the times indicated, using a radar range setting of 15 miles:

Time	Bearing	Range (mi.)
1900	035°	14.4
1906	035°	12.8
1915	035°	10.4
1924	035°	8.0
1933	035°	5.6
1941	030°	3.5
1947	015°	1.9

Required—(As determined at 1915.)

(a) Range at CPA.

(b) Time at CPA.

(c) Course of other ship.

(d) Speed of other ship.

Decision—When the range decreases to 5.0 miles, own ship will change course to the right, maintaining a speed of 20 knots, to pass the other ship with a CPA of 1.0 mile. Original course of 035° will be resumed when the other ship is broad on the port quarter.

Required—

(e) Predicted time of change of course to the right.

(f) New course for own ship.

(g) Bearing of CPA as determined at 1933[1].

(h) Predicted time at 1.0 mile CPA as determined at 1933[1].

(i) Bearing of other ship when own ship commences turn to original course.

(j) Predicted time of resuming original course.

1. We believe the Problem as published in Pub 1310 erroneously asked for 1935 times.

Answers. (a) nil; collision risk exists, (b) 1954, (c) C 035, (d) 4.0 kt, (e) 1935, (f) C 044, (g) 316, (h) 1954, (i) 269, (j) 1958

Plot target at appropriate ranges and bearings; note times. Note constant target bearing and decreasing range; therefore collision risk exists (a). Measure and calculate SRM; calculate TCPA (b). Draw 6-minute buoy trail and compare to distance between 6-minute contacts. Observe that buoy trail is longer than distance between contacts; therefore, we are overtaking another vessel of the same course (c). Calculate difference between buoy trail and distance between contacts as actual speed of target (d). Use SRM to calculate time at 5.0 nmi Turn (e). Draw 1 nmi range ring. Draw new DRM from Turn point to pass tangent to 1 nmi range ring. Parallel new DRM to convenient open area. Draw a line representing target's now known actual course and speed (can use 15-minute values to make drawing large) and position adjacent to new DRM. Draw 15-minute buoy trail and position to pivot from astern of target's 15-minute actual course and speed line. Swing buoy trail until other end is adjacent to DRM line. Read new course for own ship (f). Measure bearing to CPA (g). Measure and calculate new SRM; calculate TCPA (h). Draw radius from own ship representing 'broad on the port quarter' (i). Use SRM to calculate contact time at this bearing (j).

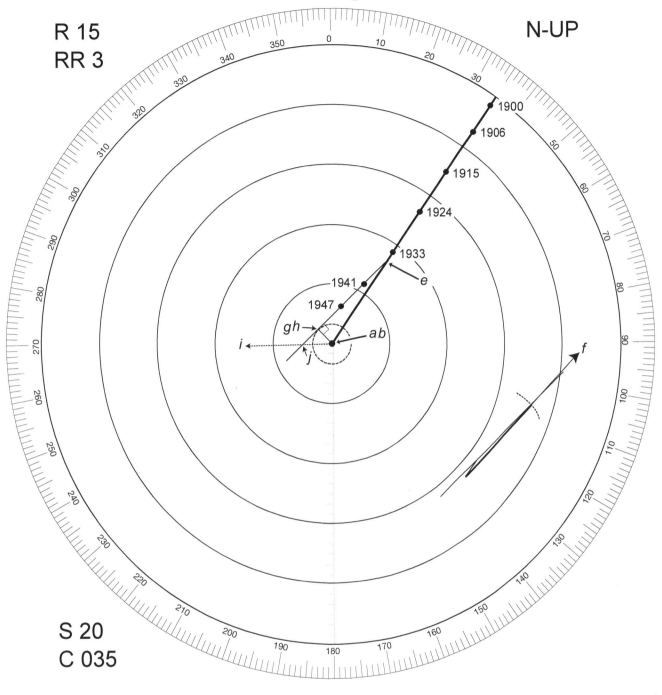

R 15
RR 3

N-UP

S 20
C 035

26. Own ship, on course 173°, speed 16.5 knots, obtains the following radar bearings and ranges at the times indicated, using a radar range setting of 20 miles:

Time	Bearing	Range (mi.)
2125.5	221°	16.0
2130	220.5°	15.0
2137.5	219°	13.2
2142	218°	12.2
2151.5	215.5°	10.0
2158	205.5°	8.3
2206	185°	6.7

Required—(As determined at 2142.)

(a) Range at CPA.

(b) Time at CPA.

(c) Predicted range other ship will be dead ahead.

(d) Predicted time of crossing ahead.

(e) Course of other ship.

(f) Speed of other ship.

Decision.—When range decreases to 10 miles own ship will change course to the right to bearing of stern of other vessel (assume 0.5° right of radar contact).

Required—

(g) Range at new CPA.

(h) Time at new CPA.

(i) Direction of new relative movement line.

(j) New relative speed.

(k) New course of own ship.

Decision—Own ship will resume original course when bearing of other vessel is the same as the original course of own ship.

Required—

(l) Predicted time of resuming original course.

(m) Distance displaced to right of original course line.

(n) Additional distance steamed in avoiding other vessel.

(o) Time lost in avoiding other vessel.

Answers. (a) 2.5 nmi, (b) 2233, (c) 3.0 nmi, (d) 2226, (e) C 120, (f) 14.7 kt, (g) 6.4 nmi, (h) 2211.5, (i) 075, (j) 23.3 kt, (k) C 216, (l) 2209, (m) 3.3 nmi, (n) 1.3 nmi, (o) less than 5 min

Plot target at appropriate ranges and bearings; note times. Draw DRM. Measure range at CPA (a). Measure and calculate SRM; calculate TCPA (b). Measure range when target is dead ahead (c). Use SRM to calculate time target dead ahead (d). Draw 14-minute buoy trail commencing 2137.5 contact. Complete RMD to determine

Radar Maneuvering Problem 26.

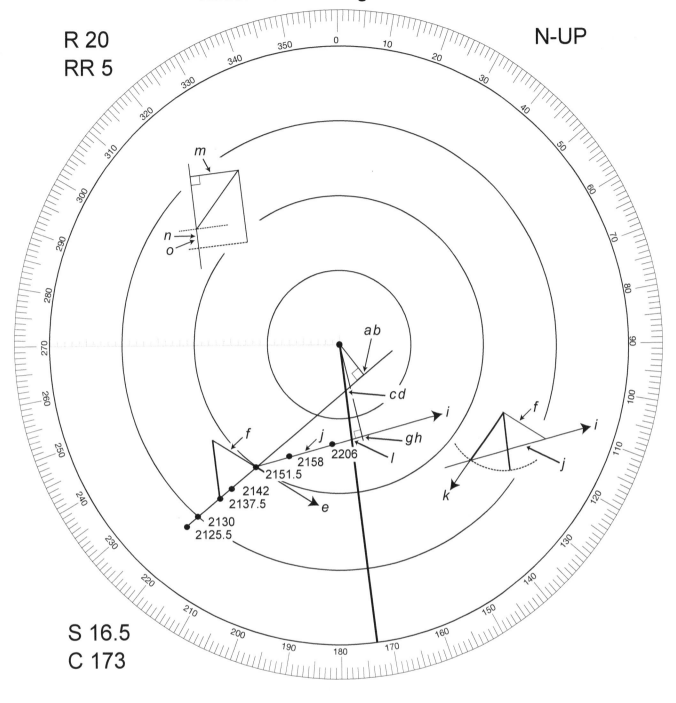

R 20
RR 5

N-UP

S 16.5
C 173

target's actual course (e) and speed (f). In a convenient open area copy 14-minute buoy trail and 14-minute target actual course and speed line. Swing buoy trail to desired new course, ie, original course + .5 degree (k). Complete RMD with new DRM (i). Measure and calculate new SRM (j). Parallel new DRM to commence 2151.5 contact. Measure range at CPA (g). Calculate new TCPA (h). Calculate time when target bears same as original course (l). In a convenient open area draw a reference line of original course. Draw another line of same course that has length = distance run between 2151.5 and beginning of turn back to original course. Separate these two lines by line = course and distance run between the two course changes. Read distance offset (m). Compare advance of alternative actions and read additional distance steamed (n). Calculate time lost avoiding in avoiding other vessel (o).

27. Own ship, on course 274°, speed 15.5 knots, obtains the following radar bearings and ranges at the times indicated, using a radar range setting of 20 miles:

Time	Bearing	Range (mi.)
0815	008°	14.4
0839	006°	10.1
0853	004°	7.6

Required—

(a) Range at CPA.

(b) Time at CPA.

(c) Course of other ship.

(d) Speed of other ship.

Decision—When the range decreases to 6.0 miles, own ship will commence action to obtain a CPA distance of 4.0 miles, with own ship crossing astern of other vessel

Required—

(e) Predicted bearing of other ship when at a range of 6.0 miles.

(f) Predicted time when other ship is at 6.0 mile range, and own ship must commence action to obtain the desired CPA of 4.0 miles.

Decision—Own ship may

(1) alter course to right and maintain speed of 15.5 knots, or

(2) reduce speed and maintain course of 274°.

Required—

(g) New course if own ship maintains speed of 15.5 knots.

(h) Predicted time when other vessel bears 274° and own ship's original course can be resumed.

(i) New speed if own ship maintains course of 274°.

(j) Predicted time when other vessel crosses ahead of own ship and original speed of 15.5 knots can be resumed.

Answers. (a) 1.1 nmi, (b) 0935, (c) C 242, (d) 20.1 kt, (e) 002, (f) 0902, (g) C 019, (h) 0916, (i) 8.3 kt, (j) 0936

Plot target at appropriate ranges and bearings; note times. Draw DRM. Measure range at CPA (a). Measure and calculate SRM; calculate TCPA (b). Draw 38-minute buoy trail; complete RMD to determine target's actual course (c) and speed (d). Use SRM to determine time at 6-mile turn point (f); measure bearing to turn point (e). Draw 4-mile range ring. Draw new DRM from turn point to pass tangent to 4-mile ring. Parallel new DRM to convenient open area. Copy target's 38-minute course and speed line to position adjacent to paralleled DRM line. Copy 38-minute buoy trail to astern end of target course and speed line and swing right to contact paralleled DRM. Read new course for own ship (g). Draw shortened buoy trail to intercept DRM line on original heading to determine new speed (i). Read new SRM values and calculate times target crosses 274 bearing (h) and (j).

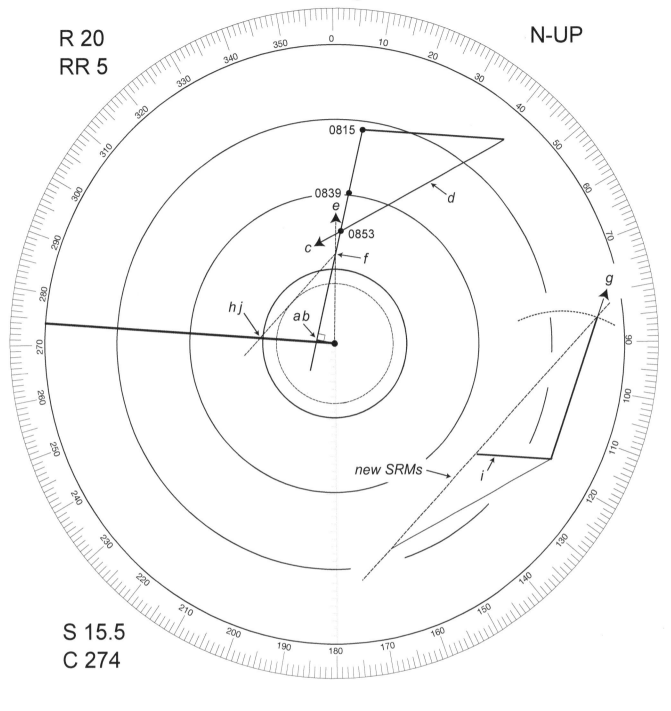

R 20
RR 5

N-UP

0815

0839

0853

e

c

d

f

g

h j

a b

new SRMs

i

S 15.5
C 274

28. Own ship, on course 052°, speed 8.5 knots, obtains the following radar bearings and ranges at the times indicated, using a radar range setting of 20 miles:

Time	Bearing	Range (mi.)
0542	052°	18.5
0544	052°	17.5
0549	052°	15.0
0550	052°	14.5

Required—

(a) Range at CPA.

(b) Time at CPA.

(c) Course of other ship.

(d) Speed of other ship.

Decision—At 0555, own ship is to alter course to right to provide a CPA distance of 2.0 miles on own ship's port side.

Required—

(e) Predicted bearing of other ship when own ship changes course.

(f) Predicted range of other ship when own ship changes course.

(g) New course for own ship.

Own ship continues to track other ship and obtains the following radar bearings and ranges at the times indicated, using a radar range setting of 20 miles:

Time	Bearing	Range (mi.)
0559	050°	10.0
0604.5	043.5°	7.4
0606.5	040°	6.5
0609	034°	5.5

Required—

(h) Course of other ship as determined at 0609.

(i) Speed of other ship as determined at 0609.

(j) Range at CPA as determined at 0609.

Answers. (a) R 0.0 nmi, (b) 0619, (c) C 232, (d) 21.5 kt, (e) 052, (f) 12.0 nmi, (g) C 086, (h) C 241, (i) 21.5 kt, (j) 3.0 nmi.

Problem 28. Plot target at appropriate ranges and bearings; note times. Draw DRM. Observe target bearing is not changing as range decreases; therefore, range at CPA is nil (a); collision risk exists. Measure and calculate SRM; calculate TCPA (b). Observe that buoy trail is less than time between contacts; therefore target is approaching on reciprocal course (c). Calculate target speed as difference between own ship speed and time between contacts (d). Use SRM to calculate range at Turn point (f). Observe that target bearing has not changed when

Radar Maneuvering Problem 28.

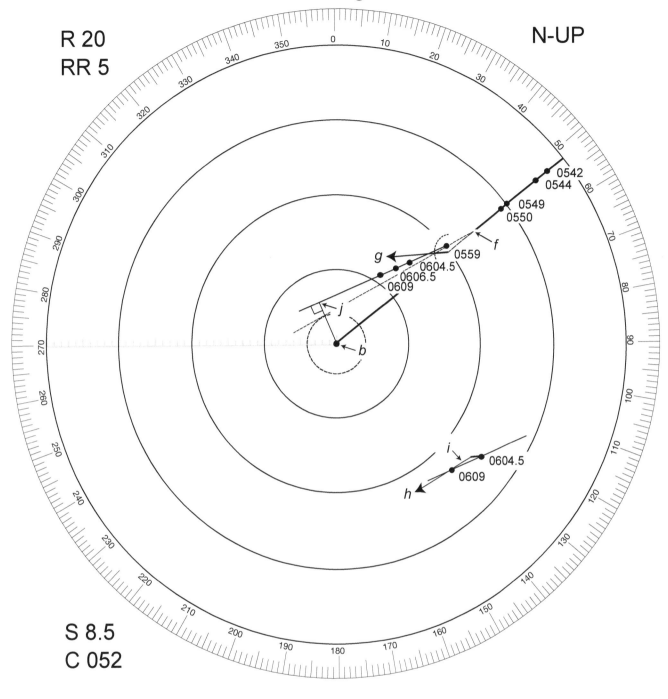

R 20
RR 5

N-UP

S 8.5
C 052

course change is initiated (e). Draw 2-mile range ring. Draw new DRM to pass tangent to 2-mile ring. Draw now known target 6-minute course and speed line commencing at Turn point. Draw 6-minute buoy trail from bottom of target line and swing to intercept new DRM line to determine new course for own ship (g). Draw a third DRM commencing 0559 contact. Measure range at CPA (j). Copy DRM and contacts to convenient working area. Draw 4.5-minute buoy trail commencing 0604.5 contact. Complete RMD to determine target new course (h) and speed (i).

RADAR PLOTTING SHEET

The sheet may be duplicated for practice with radar plotting. (4-ring sample on page 46)

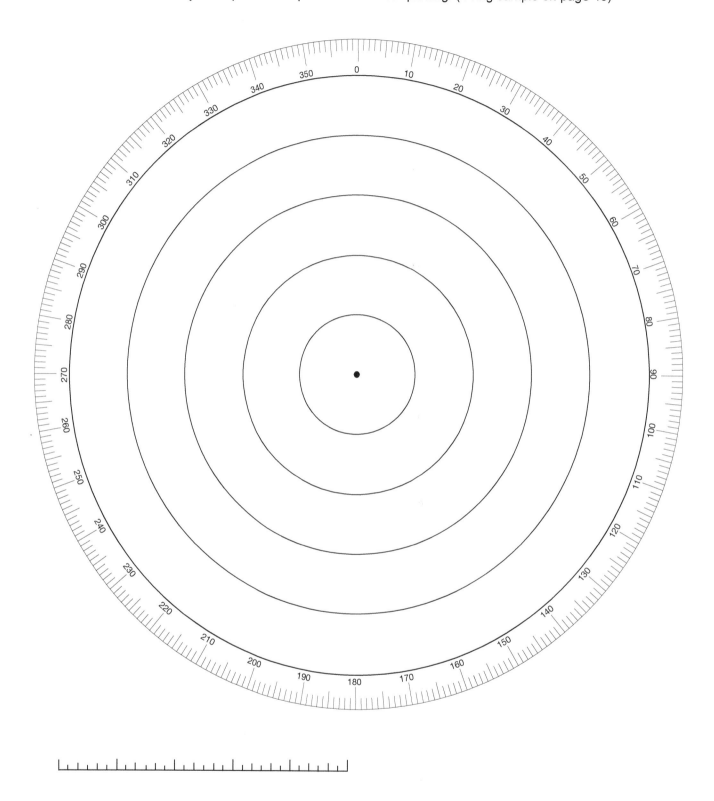

CPSIA information can be obtained
at www.ICGtesting.com
Printed in the USA
LVHW060844130221
679240LV00051B/815